CHILDREN of CHERNOBYL

RAISING HOPE FROM THE ASHES

MICHELLE CARTER and
MICHAEL J. CHRISTENSEN

Introduction by Olga Korbut

Augsburg
MINNEAPOLIS

CHILDREN OF CHERNOBYL
Raising Hope from the Ashes

Scripture texts unlesss otherwise noted are from the New Revised Standard Version Bible, copyright 1989 by the Division of Christian Education of the National
Council of the Churches of Christ in the United States of America. Used with
permission.

Cover: Lecy Design

Interior design: Virginia Aretz, Northwestern Printcrafters

Photo credits for photo section: 1A, 4A, 6B, 9A, Veronica Hernandez; 1B, 3A,
3B, 5A, 5B, 5C, 6A, 7B, 7C, 9B, 11B, Greg Schneider; 2A, Michelle Carter; 2B,
4B, 8B, 14A, 16B, Citihope; 7A, 8A, 12B, 14B, 15A, Michael Christensen; 10A,
10B, Arthur Pollack — *The Boston Herald*; 11A, 11C, 12A, 12C, Andrew Schukin;
13A, *San Francisco Examiner*; 13B; *Muskegon Chronicle* photo by John McConnico;
15B, 15C, 16A, Kristina Brendel.

Library of Congress Cataloging-in-Publication Data

Carter, Michelle
    Children of Chernobyl : raising hope from the ashes / Michelle
Carter and Michael J. Christensen : introduction by Olga Korbut.
        p.   cm.
    Includes bibliographical references.
    ISBN 0-8066-2685-2 (alk. paper) — ISBN 0-8066-2677-1
(alk. paper : pbk.)
        1. Child welfare—Belarus.   2. Radiation victims—Services for—
Belarus.   3. Church work with disaster victims—United States.
4. Chernobyl Nuclear Accident, Chernobyl', Ukraine, 1986—Health
aspects.   I. Christensen, Michael J.   II. Title.
HV799.B38C27   1993
362.1'9892'000947714—dc20   93-9757
                                                                CIP

Manufactured in the U.S.A.   AF 9-2677

97   96   95   94   93   1   2   3   4   5   6   7   8   9   10

To the children of Chernobyl
in Belarus, Ukraine, and Russia
who need a voice to tell their story,
a heart to feel their pain,
and the world to be their advocate.

CONTENTS

PART FIVE

PREFACE

"If I could do anything," writes a school child near Chernobyl, "I would eliminate radiation so I could walk in the woods again and swim." Sadly, in much of his homeland such ordinary pleasures have been lost, not only for his lifetime, but for many thousands of years to come.

Life will never be simple and uncomplicated again for two million residents of Belarus, and millions more in Ukraine and southern Russia, who continue to be exposed daily to off-the-chart levels of radiation. In Belarus alone, more than 800,000 children are at risk.

Only now, years later, are the full and terrible consequences of the 1986 Chernobyl disaster becoming known. Shortly after midnight on Saturday, April 26, 1986, routine maintenance was in progress at the Chernobyl nuclear power plant in the northeast corner of the Ukraine when an uncontrolled power surge raced through reactor No. 4, producing steam and hydrogen, which culminated in a massive explosion. A mile-high nuclear cloud hovered for ten days over vast areas of the Soviet Union and Europe, releasing its nuclear rain. Prevailing winds carried radiation containing deadly isotopes of iodine, plutonium, and cesium to the northwest, dropping seventy percent of the fallout from Chernobyl on the Ukraine's neighboring republic, Byelorussia.

In 1990, individuals within Byelorussia first found the courage to defy the Kremlin, reveal the scope of their suffering, and ask the world for help.

That was the same year that the two of us first encountered the children of Chernobyl. We took quite different paths, but arrived at the same place at almost the same time. From the capital city of Minsk to the villages surrounding Chernobyl's "Dead Zone," we have come to know the children's suffering, listened to their desperate parents, and learned from the courageous doctors who care for them without access to modern medical techniques or equipment.

While this book tells a tragic story, it also offers much hope. In partnership with other volunteers, we have learned that people who care can make an enormous difference. We have witnessed a rebirth of the spirit in Belarus as many badly needed shipments of medicine and food—accompanied by love and concern—have been hand-delivered by people of faith.

This book is also a story of East-West encounter. We found ourselves immersed in the thousand-year history and culture of the Slavs, who today carry the burden of what our Belarusan publisher Alexander Lukashuk calls the three latter-day Golgothas of Belarus: Kurapaty, where Stalin dropped hundreds of thousands of his victims into shallow graves; Khatyn, where Hitler tried genocide to end Soviet resistance; and Chernobyl, where the lives of uncounted future generations were compromised.

But remarkably, through it all we found hope and renewal as more and more sick children went into remission, as the promise of democracy and freedom found expression, and as Americans heard the story and were moved to respond with compassion.

The events reported in this book are true. As we worked to put them down on paper, we found that only minor adjustments were needed to smooth out the telling. (For example, we realized how confusing it would be to readers if we referred to each of the many persons named Natasha as "Natasha.") The story begins with the Soviet republic of Byelorussia and ends with the independent nation of Belarus. The name has changed, the scope of Western relief efforts has grown, but thousands of children are still in danger.

We invite you to read on and meet several of the children of Chernobyl. We have been able to write about only a few of them, but they speak for all the Natashas and Vovas and Svetlanas and Andreis, who remain unnamed but not unloved. They will touch your heart and your soul, as they have touched ours.

Michelle Carter
San Mateo, California

Michael J. Christensen
Drew University
Madison, New Jersey

ACKNOWLEDGMENTS

This book would not even have been conceived without the interest of Vladimir Lipsky, a popular writer and editor of children's magazines in Belarus. As director of the Byelorussian Children's Fund, he served as host for a number of our humanitarian delegations. He encouraged us to "write the story of your visits, help us understand why Americans care about the children of Chernobyl, and why you travel so far to help. Charity is a new idea to us and we want to learn from you."

He arranged for us to meet his publisher, Alexander Lukashuk of Belarus Publishers, who offered us a contract with royalties paid in rubles. We assumed our manuscript would simply be a labor of love for the people of Belarus.

However, others believed in the merit of our narrative and urged us to seek an English-language publisher. Many thanks to Paul, Sharon, and PJ Moore, Lonnie Hull, Michelle Rapkin, Mary Catherine Dean, and Clayton Carlson, who read the manuscript and encouraged us in the process.

Robert Moluf, former editorial director for Augsburg Books, was the one who challenged us to rewrite our manuscript for audiences in America and Europe. Under his wisdom and guidance, and through his editorial partnership with Lucile Allen, the story took its present form. We gratefully acknowledge and thank them and the staff at Augsburg for timely publication.

Finally, to the most important contributors to the process, both professionally and personally, our spouses—Laurie Carter and Rebecca Laird-Christensen—without whom this book would not have been written and rewritten, and our children—Robyn and David Carter and Rachel and Megan Laird-Christensen, who constantly reminded us of our purpose in tackling this project, we offer our love and thanks.

INTRODUCTION
by Olga Korbut

Children are sick and dying throughout my homeland, the former Soviet republic of Byelorussia, the result of the terrible explosion at the Chernobyl nuclear power plant in April 1986.

The radioactive poisoning of that beautiful land was terrible enough, but it was turned into an unending horror because the government failed to protect its people. Instead of ordering everyone indoors behind sealed doors and windows or calling for an immediate evacuation as soon as it happened, nothing was said for days and days. We went on with our lives, gardening in the unusually warm spring air and celebrating the end of a long, cold winter. When we were finally told about the catastrophe, it was too late. A dreadful fate had been sealed for my beautiful Byelorussia and its children.

At that time, I was living in Minsk with my husband Leonid and my son Richard, and we became victims just like everyone else. We heard the rumors and guessed at the truth. Over the years we watched friends and neighbors get mysteriously sick, but there was never any explanation.

However, being a gymnast and Olympic gold medalist allowed me a few opportunities that were denied others in Byelorussia. Although I was not free to travel routinely to the West (we were labeled *nevyezdnoi*—"unreliable, not suitable for travel abroad"), I was invited in 1989 to come to Indianapolis, Indiana, for an exhibition by the American Gymnastics Federation. Remarkably, the state granted permission for my family and me to go.

When word spread that we were going to the United States, mothers and fathers began turning up on our doorstep or calling to ask us to help their children, who were suffering from leukemia or some other form of radiation poisoning caused by the

Chernobyl disaster. I was amazed at the number of requests and the desperation of their pleas.

One father I remember well was a taxi driver whose daughter, Irina, had leukemia. He told us that there was no comprehensive pediatric hospital in Minsk that could save his daughter. "I know there is a clinic in America that saves children with leukemia. Would you take the samples of her blood with you to show the doctors there?"

I couldn't refuse but I had no idea what I could do once I was in the United States. I only said that I would try. It took nearly six months to obtain the official permission to take several hundred glass slides of children's blood samples with us to Indianapolis, but God was on our side when we got there. Our guide turned out to be a representative of the American Red Cross who knew exactly what had to be done to get the slides evaluated.

I was overwhelmed by the warm and loving reception I received in America. They remembered me after all those years. When I was told that a press conference was scheduled, I decided to use the opportunity to talk about the situation in my country. It had been years since the accident and I didn't know how much people in the United States knew—or even cared—about the effects of Chernobyl.

As passionately as I could, I told the reporters in Indianapolis how terrible life in the radiation zones had become, that the soil and the air were poisoned with radiation, and no one knew how many centuries would pass before they would be clean again. I told them there was no escape for the children. They could no longer walk in the woods and pick apples and mushrooms or swim in the rivers or behave like normal children.

Then I held up the medical files and blood sample slides I had brought with me and told the stories of the children who were dying because there were no medicines or treatments or medical facilities to save them.

When I left the room, I wasn't sure that the reporters believed me, but they must have. They wrote wonderful news stories and the people of Indianapolis opened their hearts. A local hospital called the next day and offered to treat some of the leukemia patients. Another group formed a foundation to help.

As soon as we returned to Minsk, I went to work to get the taxi driver's daughter, Irina, to that hospital in Indiana for treatment. I knew it wouldn't be easy, but even I was surprised; it took five long months to obtain the proper documents. By the time they were in my hands, the child had died.

I raged at the system that would let sick children die when help was available, but nothing changed.

At the same time, I was beginning to worry about my son Richard's health. Finally, Leonid and I reached a painful decision: we would send him to live with friends in New Jersey. It was difficult, but we were now able to travel back and forth more easily and we were able to see him often. Then, in 1990, we were able to immigrate ourselves and reunite our family in the United States.

Soon I was giving gymnastics exhibitions on a fairly regular basis, and my manager was based in Seattle, Washington. While visiting there, we learned about the Fred Hutchinson Cancer Research Center and the wonderful work they do there with bone-marrow transplants. I was particularly interested because such transplants are the only hope for children with leukemia who don't respond to chemotherapy. They had everything necessary to treat children of Chernobyl, and they were willing to help.

That's when I decided to establish the Olga Korbut Foundation—in conjunction with the Hutchinson Center. I would use my name to raise money to bring sick Byelorussian children to Seattle and to pay for the highly effective but very expensive bone-marrow treatments at the Hutchinson Center. It was there in Seattle that I first learned about Citihope and its efforts for the children of Chernobyl.

The children of Chernobyl are the sons and daughters of my homeland. They were victims of a government that didn't care enough to protect them, to tell them to stay indoors in the days after the explosion, to tell us parents how we could protect our own children. And now they are sick. Many have leukemia and thyroid cancer; still others are weak and listless and cannot fight off a cold or a runny nose. Other children have nosebleeds or stomachaches or are losing their hair, and no one can explain why.

Now the Soviet Union has disappeared, but its problems have only grown larger. There is little money to buy the medicine or build the hospitals that will save these children, so mothers must sit by their beds and watch them die.

The radioactive contamination that is the legacy of Chernobyl will be present in the now independent republic of Belarus for generations to come, but it must no longer be a secret. The world must know about the children of Chernobyl and their critical needs.

I am pleased to introduce this book to you, and I encourage you to read on. These children will tear at your heart, but you must know them and respond. Our planet is very small, and they are your children, too.

Olga Korbut
Atlanta, Georgia
February 1993

Belarus (Formerly Byelorussia) and Surrounding Countries

Areas Contaminated
by Radiation from Chernobyl

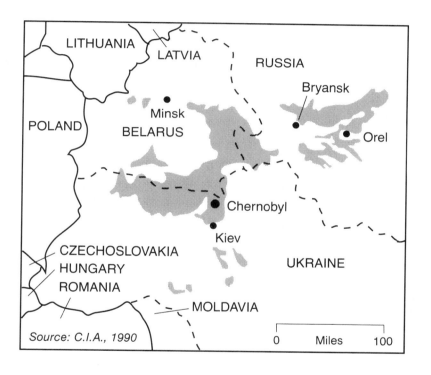

LITHUANIA
LATVIA
RUSSIA
Bryansk
Minsk
POLAND
BELARUS
Orel
Chernobyl
Kiev
CZECHOSLOVAKIA
HUNGARY
UKRAINE
ROMANIA
MOLDAVIA

Source: C.I.A., 1990

0   Miles   100

TIME LINE

April 26, 1986 — Nuclear Reactor No. 4 at Chernobyl in the northeast corner of the Ukraine explodes shortly after midnight, spewing radioactive isotopes into the atmosphere. Prevailing winds carry them in a north-northwesterly direction over the neighboring republic of Byelorussia, where seventy percent of the fallout settles and puts 2.2 million people and 800,000 children at risk. The accident produces ninety times the radioactive fallout of the atomic bomb dropped on Hiroshima and one million times the emissions of the accident at Three Mile Island.

April 28, 1986 — Elevated levels of atmospheric radiation are reported throughout Europe and Scandinavia, particularly in Sweden, and international speculation focuses on a nuclear accident in the Soviet Union. The Kremlin is silent.

May 1, 1986 — May Day celebrations take place as planned throughout Byelorussia and the Ukraine on an unusually warm spring day.

May 6, 1986 — The first evacuation order for people living within thirty kilometers of the Chernobyl nuclear reactor is issued — ten days after the explosion.

May 14, 1986 — Soviet President Mikhail Gorbachev addresses the nation about the Chernobyl accident: "Today, we can say that because of the effective measures taken, the worst is behind us."

July 1989 — The Russian Orthodox Church makes an appeal for humanitarian assistance for the victims of Chernobyl through Orthodox churches worldwide.

Summer 1990 — Actor Paul Newman's Hole-in-the-Wall Gang Camp, the American Cancer Society of Michigan, and the YMCA Family Camp of Michigan invite children of Chernobyl with cancer to their summer camps.

Summer/Fall 1990 — Pyotr Kravchanka, Byelorussian foreign minister, makes two appeals to the General Assembly of the

United Nations for a worldwide response to the growing medical and environmental crisis in Byelorussia.

August 1990—Citihope International learns about the critical needs of the children of Chernobyl through newspaper articles and meets with the Byelorussian ambassador in New York. A fact-finding team makes Citihope's first delivery of medicines in September.

November 1990—A United Church of Christ delegation from northern California visits Minsk, the capital of Byelorussia; a second Citihope team arrives with 1,800 vials of methotrexate.

January 1991—A third Citihope delegation participates in the first officially sanctioned Orthodox Christmas celebration in the Soviet Union in seventy years.

April 1991—Sharon Moore accompanies another shipment of supplies to Minsk and performs at concert commemorating the fifth anniversary of the explosion at Chernobyl.

June 1991—A fourth Citihope delegation, with medical personnel and supplies, travels to Byelorussia.

Summer 1991—Project Fresh Air provides radiation-free rest and recreation for thirty-five children of Chernobyl in California and New York.

August 19, 1991—Soviet President Mikhail Gorbachev is toppled from power by a coup. The coup is broken in a matter of days by pro-democracy demonstrations led by Boris Yeltsin.

August 25, 1991—The conservative republic of Byelorussia declares its "political and economic" independence from the Union of Soviet Socialist Republics after a unanimous vote by its Communist-dominated parliament.

September 19, 1991—The republic of Byelorussia changes its name to Belarus and adopts a pre-communist flag that had recently been the symbol of the Popular Front.

October 29, 1991—The Ukrainian Parliament votes to close the Chernobyl nuclear power plant, which produces about three percent of the republic's energy, within two years. This action follows a three-and-a-half hour fire on October 11 in which the roof of one of the two remaining reactors was destroyed.

December 1991—The Supreme Rada parliament of Ukraine issues a report after a sweeping probe of the Chernobyl disaster and accuses the Communist leaders of 1986, including Gorbachev, of "a massive criminal cover-up that led to thousands of deaths." The Soviet leadership, the report says, reacted with "a total lie, falsehoods, cover-up and concealment . . . .Everyone in the upper echelons of power knew everything" within twelve hours of the explosion and went ahead to convince people that nothing was wrong.

December 6-8, 1991 — The Union of Soviet Socialist Republics is officially dissolved, to be effective on the last day of the year, at a meeting in a woods outside Minsk. It is replaced by an informal confederation of eleven of the former Soviet republics, called the Commonwealth of Independent States.

December 25, 1991 — President Mikhail Gorbachev resigns.

December 31, 1991 — A Citihope group launches a relief mission during the final days of the Soviet Union.

June 1992 — Operation *Nadezhda* Express brings 7.7 million pounds of U.S. Department of Agriculture foodstuffs, including flour, rice, and cooking oil, to Belarus for distribution at six sites within the republic. Sixty-three American volunteers, organized through Citihope International, hand out the food to the needy.

January 1993 — The Bread and Butter Initiative, an ambitious partnership between the U.S. Department of Agriculture, the Belarusan Ministry of Humanitarian Assistance, and Citihope International, is launched to provide emergency relief and economic aid over a three-year period, using American volunteers in the manner of Operation *Nadezhda* Express.

April 26, 1993 — The *Dukhovnost* (Spirituality) Center, sponsored by World Vision, opens in Gomel, Belarus, for the promotion of mental health and pastoral care among the children of Chernobyl and their families.

PART ONE

Round-faced and saucer-eyed, a towheaded girl sat on a small painted stool at the edge of a dusty road, stretching from her perch to look up the lane every minute or two. She tapped her feet and fidgeted impatiently with her red scarf as nine-year-olds will do. Eyes of a surreal, pale blue punctuated her wide Slavic face, and excitement flushed her cheeks.

Natasha Ptushko was waiting for a parade!

It was May Day 1986—the most important day of the year in her small town in the southeastern corner of Byelorussia. Every city and town in the Soviet Union would host a parade that day. Natasha and her sister would soon join all the other Young Pioneers in her town as they marched down the road on their way to the square.

Beads of perspiration dotted Natasha's turned-up nose. Her mama said this was the hottest spring she ever remembered. They had all been outside in the sun—and the unseasonably warm rain—every day for the past week. Natasha poked the toe of her shoe at a single purple lupine that had popped out in the premature warmth of the early spring.

The sounds of children giggling and singing at a distance broke the stillness and Natasha bounded out of her chair. "Natalia, they're coming! They're coming right now!" she shouted. "Hurry up or they might miss us!"

The blue, carved wooden gate swung open through the overgrown grass and an exact duplicate of Natasha, in the same red scarf, appeared at the roadside. They were twins, identical blonde twins with the same eerie, nearly opaque blue eyes.

The dust, stirred by dozens of small, stomping feet, billowed up from the road and set Natasha coughing as she grabbed her sister's hand and pulled her into the group. It was May Day and the warmest day of the year. Nothing could spoil Natasha's parade.

1

"THE WATERS
TURNED BITTER ... "

FOUR AND A HALF YEARS later, Pyotr K. Kravchanka
stood at the podium of the General Assembly of the United Nations, gazed out at the delegates adjusting their earphones to
pick up the translation of his words, and breathed deeply.

A tall, rather refined-looking man of about forty-five, the
Byelorussian minister of foreign affairs gathered his courage,
summoning all the personal strength he possessed, before he
began to pronounce the words that would condemn his own government.

A lifelong Communist and minister of one of the few truly
tranquil Soviet republics, he was about to do the unthinkable—
perhaps the unforgivable. He would defy the public policy of the
Politburo in Moscow and plead for help on this world stage.

"There are few nations to which history has been so cruel,"
he began in his compelling and literary style. "For centuries our
territories, being a sort of crossroads of Europe, were not spared
a single invasion, campaign, or aggression. Wars, mass deaths,
epidemics have reduced with shocking consistency and persistence the Byelorussian population by a quarter or a half, at least
once every century. From the middle of the seventeenth century
to the end of the eighteenth century, plague reduced it by half.
At the beginning of the nineteenth century every fourth inhab-
itant died. During World War I, every fifth Byelorussian died. . . .
And the whole world knows that, in the holocaust of the Second

3

World War, one out of four inhabitants of our republic was killed.

"And now there is a new ordeal—Chernobyl, the Calvary of the twentieth century for the Byelorussian people."

Kravchanka drew a poignant picture of the 1986 catastrophe. "Seventy percent of the radioisotopes from Chernobyl have landed on the republic. One-third of the territory is in their grip, and 2.2 million people, including 800,000 children, have become innocent victims of Chernobyl."

He described a "radiation desert of uninhabited reserves of many hundreds of thousands of acres surrounded by barbed wire. . . . According to the most modest of estimates, it will not be possible to live there for thousands of years to come."

"One should state with utmost openness and bitterness that only now, after four and a half years, are we breaking through the wall of indifference, silence, lack of understanding. And we are guilty of much of this ourselves. History has yet to bring in a moral verdict on those who . . . (have) been hiding from the people the real truth about the consequences of the accident.

"The truth is, the explosion produced 90 times the radioactive fallout of the atomic bomb dropped on Hiroshima and one million times the emissions of the nuclear accident at Three Mile Island. With its long-term effects, Chernobyl has turned out to be the gravest technological disaster of modern times."

Then, remarkably, this foreign minister of an officially atheistic Soviet republic quoted from the last book of the Bible, the Revelation to John:

" ' . . . A great star, blazing like a torch, fell from the sky on a third of the rivers and on the springs of water—the name of the star is Wormwood. A third of the waters turned bitter, and many people died from waters that had become bitter.'

"The Slavic languages have the word *chernobyl*, which means 'wormwood' or 'bitter grass,' " he continued. "I am devoid of fatalism, but who can fail to be moved by the mournful and tragic lines from the book of Revelation?

"I am firmly convinced that the world community cannot step into the twenty-first century with a clear conscience without saving the children of Chernobyl. Our people believe and trust that other people of good will, fellow dwellers of our common home which is the planet earth, will not leave (us) alone facing this awful disaster."

As he ended his passionate plea, Kravchanka recalled lines from a 1926 poem by Byelorussian poet Ouladzimir Dubouka:

> "Oh, Belarus, my wild rose,
> A green leaf, a red flower,
> Neither whirlwind will ever bind you
> Nor wormwood will ever cover."

❄ ❄ ❄

PAUL MOORE, a Nazarene minister serving an Episcopal parish, sat in his cramped office in the landmark Lamb's Club building near New York City's Times Square. He and his wife Sharon and a small staff ran Citihope, an urban ministry with a weekly radio broadcast and an available network of supporters and volunteers.

Paul tapped the temple of his silver-gray head with the folded pages of the *New York Times* and mused as he stretched his six-foot, four-inch frame behind his desk. A slight smile broke across his face.

Without changing his relaxed pose, Paul called out to the next room. "Michael, are you in there? You've got to read this. There's something I think we should do."

A younger man in his thirties with a trimmed, light brown beard leaned into the room with a telephone receiver cradled on his shoulder and pointed to the phone.

Paul ignored Michael's gesture, waved the newspaper at him, and continued. "There's something here that could be important, really important."

After freeing himself from his phone conversation, Michael came in and read the article that Paul had folded to the front of the paper. The story described the efforts of Kravchanka and the permanent mission of the Byelorussian SSR to the United States to generate worldwide support for the children of Chernobyl. It outlined the programs of a government-sponsored organization, the Byelorussian Children's Fund, to save those children. The article also profiled the efforts of the YMCA in Lansing, Michigan, to bring Byelorussian children to the United States.

Paul watched Michael's face while he read. Michael Christensen was a fellow Nazarene minister who, like him, worked outside that denomination. Paul first met him in the early seventies, during the so-called Jesus Movement, when Paul was a long-haired street preacher and founding pastor of the Lamb's Church and mission in Times Square. Michael was a disillusioned college student on a spiritual quest, using his summers and semester breaks to test the waters of urban ministry.

Michael was attracted to Paul's personal charisma, entrepreneurial spirit, and passion for the poor; Paul appreciated Michael's intellect and winning way with people. They shared a common interest in working to make the church relevant and responsive to the culture and to human need.

Paul was quick to hire Michael as his associate, and mentored him in the art of getting things done. Less intense, more subtle and thorough, Michael balanced Paul's visionary flights of grandeur. They had found a way to work together, as friends and colleagues, for nearly twenty years.

When Paul started Citihope in the early eighties, Michael had already moved to San Francisco, first to found and direct Golden Gate Community Church and, later, to direct the United Methodist AIDS Project. However, he continued to serve on Paul's board of directors and worked as a special consultant to Citihope. That morning in the fall of 1990, Michael was in New York for a board meeting and consultation.

"So what do you think?" Paul asked as Michael put the newspaper down on the desk. "It fits everything we've been talking about doing in Eastern Europe."

The incredible political events of the previous year—the fall of the Berlin Wall and the freedom movements in Poland, Czechoslovakia, Hungary, and Romania—had captured Paul's interest, and he was curious to see if Citihope's highly successful technique of connecting existing needs with available resources in urban America could work on an international basis.

Until that morning, he had focused on three situations: Berlin and its reunification, which would be celebrated that October; poor and homeless families in and around Warsaw, Poland; and orphans and children with AIDS in Bucharest, Romania.

"So, you want to add the children of Chernobyl in Byelorussia to the list?" Michael wasn't even sure how to pronounce the name of the Soviet republic or how to find it on the map.

"Well, I want you to check it out. Find out all you can right away. Who knows, maybe we could take Christmas presents to those kids. They probably don't even know who Jesus is after seventy years of atheism. I like the idea of bringing Christmas to the children of Chernobyl." He leaned back again in his chair, adding, "I like it a lot. Call the United Nations. See what you can find out."

Michael was a natural networker and researcher who had learned from Paul how to tap resources and make things happen, often by telephone. It wasn't long before he had spoken to Alexander Vasiliev, counselor to the Byelorussian ambassador, to arrange an appointment at the Soviet consulate on East 62nd Street in New York. By then, he knew how to pronounce *BYEH-lo-rush-a.*

A few days later, Paul and Michael, both wearing their clerical collars, found themselves in a consulate reception room seated on a couch next to an exquisite copper sculpture of St. George slaying the dragon. The piece was disarming, in that the dragon of this St. George, the patron saint of czarist Russia, was actually a cluster of nuclear missiles. A small plaque identified it as a smaller version of a sculpture that had been presented to the United Nations by the Soviet Union in the spirit of peace and disarmament.

Vasiliev, a small man of wiry build and reserved expression, entered through a door across the room, walked over, and shook hands. Michael noticed that Paul, who usually dominated these situations with some relaxed conversation, was tense and uncomfortable.

"That sculpture is really spectacular. It captures the essence of what both our countries are trying to achieve as we take these steps toward disarmament," Paul said in a formal, somewhat stilted, tone.

"But there's a certain irony for me to walk into this building and find St. George in the room, since I'm a minister at St. George's Episcopal Church here in New York City. You've made me feel right at home."

The comment reduced the level of tension in the air and Paul went on to introduce Michael. "Michael's as eager as I am to find a way for Citihope to help the children of Chernobyl. Do you think there's any way we could come to Byelorussia this Christmas to bring presents to children in the hospitals and orphanages in the spirit of the Christ child?"

"You mean, you want to celebrate Christmas with the children on the traditional Russian Orthodox day of observance — January 7?" Vasiliev asked.

Paul and Michael nodded, holding their breath in anticipation of Vasiliev's reaction to what they feared might be too outrageous a suggestion.

The diplomat didn't speak for a moment or two; he seemed to be deeply touched by the idea. "Christmas in Byelorussia is called *Rozhestvo*. I remember celebrating the holiday as a child, giving and receiving gifts. Those are very wonderful memories for me."

Tears were brimming in Vasiliev's eyes as he struggled with his composure.

"It's possible, perhaps, in the spirit of *glasnost* and the current political climate in the Soviet Union, that it might be permitted to have some official acknowledgment of *Rozhestvo*." Vasiliev now was clearly warming to the idea.

"I will speak to the ambassador immediately and see what I can arrange. Of course, you know that Christmas gifts would be a most welcome gesture, but perhaps you might not know that these children of Chernobyl are dying right now, this very day. What they really need is *medicamentee*—medicine and medical supplies—and soon, very, very soon.

"Could you go right away, perhaps this month, to meet the children? The idea of Christmas is wonderful, but our children are dying today. Many will not live until Christmas."

The answer was, "Yes, of course, we'll go," although Paul and Michael hadn't a clue as to how to begin.

"First," Vasiliev explained, "you must have an official invitation. I think it would be best—since you are churchmen—if Philaret, the metropolitan of Minsk, issued the invitation. We will arrange it."

ONCE THEY WERE on the street and headed back to the Lamb's Building, Michael voiced the question that was on both their minds. "Don't you think that's strange? The Byelorussian ambassador to the United Nations can just pick up the phone, call the head of the Russian Orthodox Church, and suggest that he invite a couple of American ministers he has never met to come to Byelorussia as his official guests?"

"I can't imagine that happening in the United States," Paul shouted back as they picked their way through the lunchtime crowd on 67th Street. "But it's real clear that the situation there is desperate.

"So, how fast do you think we can pull all this together?" Paul asked.

Michael shrugged his shoulders as he stuffed his hands in his pockets and leaned into the brisk, fall wind. "Maybe a month."

❊ ❊ ❊

TEN DAYS LATER, Paul, Michael, and Paul's seventeen-year-old son Paul Jr., known as PJ, launched Project Open Door, a twelve-day pilgrimage to Berlin, Warsaw, and Minsk to assess the dimensions of the need in each city and decide where Citihope's outreach program could be most effective.

After a cultural feast in the soon-to-be-reunited Berlin, the three men caught the Berlin-Warsaw-Minsk Express for their first venture behind the Iron Curtain. They had trouble finding cabin No. 11. It took several trips, weaving their way through throngs of passengers speaking a variety of European languages, before they managed to load the ten bulky bags of medicine, medical supplies, and toys donated by Citihope's radio audience.

Once they settled in and endured a two-hour delay in East Berlin, the train started its twelve-hour journey to Warsaw. The Americans stared in wonder at the passing countryside of East Germany—the Germany of Hegel, Schuller, and Einstein, the Germany hidden from the West for thirty years. The excitement of rediscovery kept them alert and questioning late into the evening.

"I wonder what's in store for these people now that communism has failed them," Michael said. "What's going to happen to this quiet countryside and its simple, communal culture once Western influence takes over? You know, there's a dark side to capitalism, privatization, and democracy. I wonder if they have even a hint of what's to come.

"Eastern Europe, in correcting the abuses of socialism," he continued, "may go too far in embracing Western ways and means."

"So, what's wrong with Western ways?" PJ shot back in defense of American capitalism. Paul Jr., a tall, lanky, still ungainly teenager with close-cropped dark hair, grew up in the Reagan years and attended a military high school. Michael had been in high school in the late sixties; his opinions were formed by the Vietnam War and the politics of the Woodstock era. Their ideas often clashed, making for loud and lively entertainment on the night train.

"Neither socialism nor capitalism equitably serves the people," Michael said. "Both are about power and domination. Both systems tend to serve the self-interests of the elite and exploit the masses. The pendulum swings back and forth but seldom finds the right balance."

"Hey, America has produced the greatest culture and political system on earth," PJ argued with conviction. "Democracy is winning around the world. One by one, communist governments are crumbling."

"No question that communism has failed. It's been a seventy-year exercise in futility for these people." Michael arranged some pillows behind his head and stretched his legs on the banquette in the cramped cabin. "I'm just not sure that whatever political system replaces it is going to offer them a much better life.

"West Germany's a wealthy country, perhaps the wealthiest in Europe, but even the West Germans may not be able to afford the huge amount of money it's going to take to bring East Germany up to speed in a relatively short period of time.

"Are the West Germans willing to make the sacrifices that will be needed? How long will the East Germans be willing to be second-class citizens in their own country?"

"None of that should matter at all," PJ said as he slumped down with his elbows on the little table under the cabin window. "They're free. That's all that counts; that's all they should care about."

"Yes, they're free, but free to be what? Free to do what?" Michael's question went unanswered as PJ pulled the headset to his Walkman down over his ears and snapped a cassette into place. The hum of muted rock music dampened any further attempt at conversation.

Michael and Paul wandered into the hallway, leaving PJ to his music. A friendly group of Poles in the next cabin offered them some food. They accepted with enthusiasm, in an effort to be sociable, but once they were alone again they agreed that what they had eaten was fairly awful. The cigarette smoke in the hallways was thick and, Michael observed somewhat self-righteously, everyone drank far too much.

In Warsaw, they booked a first-class sleeper car for twenty-one dollars for the ten-hour, overnight ride to Minsk, the capital of Byelorussia. The ride was comfortable, and the imagery reminiscent of old Paul Konreid movies: steam trains pulling into bustling, fogged-in stations at midnight, spies trading secrets, and strangers meeting and falling in love.

Paul leaned back against the far side of the compartment, took out his journal, and began to write. *"The smell of smoke from a coal-fired, potbelly stove fills the air. Steam hisses out of brake cylinders; carriage wheels squeak, swaying gently out of sight, but never out of memory. . . .*

"A train pulls up next to us. Every window in each green car frames a different mystery. Fleeting impressions of a monk-like character, a Rasputin, appearing then disappearing, dissolving, out of focus. . . ."

Frequent stops offered the travelers time to reflect and record their feelings about entering, for the very first time, the Union of Soviet Socialist Republics—the Evil Empire that all three had been urged to fear. To enter by train at night offered an unexpected element of intrigue and promise of adventure.

"Nothing can prepare a child of the Cold War for that first encounter with the USSR. Nothing," Paul continued in his journal. *"As we slowly approached the border, a desolate 'no-man's land' appeared. The Polish countryside had been picturesque with quaint old houses, wide open fields with entire families harvesting the fall crops of potatoes, squash and pumpkins. . . .*

"Now appeared a dirty, vacant space, a curvy, jagged line defined by barbed wire, a deep ditch, and then a wide swath of freshly rototilled soil; now guard posts, a wall and finally, as we lurched to a stop, huge red Cyrillic letters CCCP (USSR). No 'Welcome to our country!' Just a swarm of Red Army soldiers, some with weapons, some with briefcases, converging on our train. Young men in olive drab uniforms trimmed in red with that threatening hammer and sickle."

Paul Sr. was perhaps the most eager for an adventure; Paul Jr. the most cautious and worried. They all expected hassles from the border patrol about their medical cargo and mission.

Michael, remembering how his family had reacted to the Cuban Missile Crisis of 1962 by creating a bomb shelter in the basement of their home, nervously kept rearranging the ten bags that filled the compartment.

"I wish I had some idea how concerned the border guards are going to be about all these drugs," he said. He was prepared to produce an official letter of introduction and a detailed list of their contents.

"I'm sure the communists are going to want to know what three Americans, two of them clergymen, are up to," Paul added. "It would be nice to know if they are going to put uniformed officers on us or KGB in plainclothes."

"They're going to be watching every move we make and we'd better be careful," PJ interjected. "Communists can't be trusted. They want to control the world! My friends at school think I'm nuts for making this trip across enemy lines."

"Even Ronald Reagan said, 'Trust but verify,' " Michael replied. "Let's suspend judgment and see what we find."

"You can trust if you want to, but I'm keeping my eyes wide open." PJ was getting more agitated. "I'm not even sure it's right to jump in here with bags of medicine and gifts to bail them out of a mess that they got themselves into. I'm not sure this is a very American thing to do—if you really think about it."

Finally the train jerked to a stop at the station in Brest, the first city inside Byelorussia and the Soviet Union. At least thirty soldiers, most of them looking like teenagers, stood on the platform ready to board the train. Michael raised his camera to capture the sight when a Polish passenger grabbed his arm and whispered a hoarse, "Not now."

While they waited, Paul picked up his pen again. *"Why am I so agitated?"* he wrote. *"I think it's the result of nearly 50 years of propaganda from both sides. It's* Life *magazine; it's black and white Huntley and Brinkley TV news. It's Khrushchev pounding his shoe on the United Nations podium, shouting 'We'll bury you!' It's a hundred alarms going off in grammar school and our pressing our faces against the wall and cupping our necks with our hands—because the Russians were coming. . . .*

"How does one make peace with himself, with the 'enemy,' the Evil Empire, after 40 years or more of being at war? Maybe the border crossing is a personal pursuit of all that remains in my life of 48 years that threatens me, that I perceive has the power to destroy me. . . .

"Maybe this is a passage to freedom and healing and release. . . . Maybe."

The soldiers boarded the train in pairs and began entering compartments to check documents and luggage. The three men anxiously rehearsed their story. Questions arose: How would they explain the medicine, the Bibles, the toys? Would the video camera be confiscated? Would a bribe be expected? Would they be delayed? Accompanied? Arrested? Deported?

2

"I DON'T WANT MY
CHILDREN TO DIE!"

WHEN THE OFFICIAL knock came at their door, Michael and Paul had put on their clerical shirts and collars. The soldier was unimpressed and a little bored, neither stern nor friendly. *"Documentee?"* he said as he held out his hand for their papers.

Paul, who was fumbling for his passport, rattled on and on, explaining Citihope's mission of mercy for the children of Chernobyl, adding that they were special guests of the Byelorussian Children's Fund and Philaret, the metropolitan of Minsk.

Without uttering another word, the soldier examined their passports and visas, compared each face to its corresponding photograph, handed them currency declaration forms—and left.

Paul was astounded. "That's it? Nothing? *Nada?*"

"He couldn't have cared less about us," Michael said as he fell back onto the banquette, feeling a little disappointed. A bit of adventure would have been fun. "They must have Westerners in and out of here all the time. The least he could have done was look in one or two of the bags. We could have been smuggling drugs in here and they wouldn't even have known."

PJ was sure it was part of the plot. "They just want you to think they aren't interested. Now you'll be careless and make a mistake. That's just what they want you to do."

Word spread that the train would be in Brest for at least two hours while the wheels on the cars were switched to bigger ones that could accommodate the different gauge of train track in the

Soviet Union. Larger train wheels seemed like a peculiar way to define life under the hammer and sickle.

"Well, if we're going to be here for two hours, let's get off and see what a Soviet city is like," Paul suggested. "Let's see what Gorbachev's *glasnost* really means."

"According to the guidebook, Brest is an historic city," Michael added. "This is where the Germans launched their invasion of the Soviet Union in World War II. I'm game for some exploring."

"You'd be crazy to go wandering around in some strange communist city. I'm staying right here." PJ was hunkering down in the stuffy train cabin. "And I don't mind telling you I think you're making a big mistake. You don't have any idea what could happen out there. You don't seem to understand that this is a dangerous place and you can't just go strolling around like sightseers."

The more distrustful PJ grew, the bolder and more intent Paul and Michael became on having their adventure. In the end, Paul Jr. stayed on board to guard the goods while Michael and Paul went exploring.

They tried to imagine Brest during the war as the Nazis swept through Poland into Byelorussia. The border station appeared to be an old fortress, with surrounding buildings in ruin. Many of the walls were pockmarked by bullets or shell fragments. They found an old watchtower with a giant clock, and the two decided to take a chance with the camera. They took turns posing beneath the clock at dusk.

They poked into shops and storefronts that were open but mostly empty of goods. Inside one shop, men and women stood in line to buy cubes of animal fat, eggs, and bread; others were queuing to buy lottery tickets of some kind. A few street people, perhaps mentally disturbed, sat alone on benches or dug into garbage cans in dark alleys near the station; a pair of young lovers strolled past the Americans and smiled.

The two men were hungry, but nothing looked appealing, especially not the pitted, golf-ball-size apples a hunched old woman tried to sell them. As the afternoon turned to evening, Paul and Michael still had not figured out what there was to love or hate about Brest. It was certainly not as threatening as they expected. For the most part, they were ignored. No KGB agents lurked as they walked around, and no Red Army soldiers confiscated their cameras.

Returning to the train just two minutes before departure, the two men banged on the compartment door and found PJ bouncing to the beat of American rock ("Christian rock," he had reminded them) through his headphones.

Michael tapped on the Walkman. "Is this how you insulate yourself from culture shock or, heaven forbid, perhaps a new idea?"

PJ pulled off the headset. "You think you're so brave, but really you're just being naive. I was afraid you weren't going to get back here in time. The train has moved several times while you were gone. But you weren't even thinking about the possibility that I would be stuck on this train by myself in some God-forsaken communist country. This place gives me the creeps.

"The only reason I came on this trip is I was asked to come and I wanted to help the children. Now I'm not so sure it was a good idea."

Paul and Michael were too tired to argue any further, so they ignored their growling stomachs, pulled out the seats that made into bunks, and tried to get some sleep.

It was 1:30 A.M. when the train arrived at the lighted station in Minsk. Cars and buses arrived and departed, but no one was there to meet them, not even a KGB agent to follow them to their hotel. "So, what do we do now?" Michael asked.

"You guys watch the luggage. I'll find a porter and a cab," Paul said.

A few minutes later, he returned with a short, toothless old man with bright eyes who was eager to help the sleepy, somewhat disoriented Americans. "I found an angel," Paul boasted.

The old man hailed a cab and the driver paid the porter a couple of rubles. He then charged the Americans a couple of dollars. Once they were in the cab, the driver offered to exchange rubles for dollars at an incredibly advantageous rate, but the newcomers were not ready to barter. "Just take us to the Intourist hotel," Paul told him. They were wary and not prepared to be "taken" within minutes of arriving in Minsk.

When the cab pulled up to the Hotel Belarus, an attractive Intourist representative came out to meet the Americans. She was quick to pay the driver in rubles and chided the Americans about holding onto their dollars. The driver, quite obviously disappointed, promised to come back in the morning to exchange their money.

Exhausted, the three checked in and collapsed in their room until dawn.

The next morning, they wandered through the Hotel Belarus. This modern, postwar skyscraper with its sweeping marble staircase was relatively quiet early in October. The women who mop the marble floors or wipe down the staircase hour after hour no longer had to deal with guests tracking mud where they had just cleaned. In the spartan rooms, the double-paned windows and terrace doors—open in the summer to admit breezes off the Svisloch River—had been shut and sealed to keep out the numbing cold to come.

But from the twenty-second floor the three men could see a neat and orderly city of modern buildings and broad avenues. Wide sidewalks wound along the riverside and a bridge arched from one grassy bank to the other. The office buildings lining Masherova Prospekt formed the distant skyline.

" 'Intourist hotels in the Soviet Union are an oasis of luxury amid the poverty all around,' " Michael read out loud to Paul and PJ from his guidebook. " 'Luxury' seems to be overstating things a bit, wouldn't you say?"

But the Belarus was pleasant enough, even if the Americans found the beds too short, the furnishings sparse, and the food bland and suspect.

Michael had been warned not to drink the water in the Soviet Union and to avoid eating beef and dairy products in Byelorussia, which may have been contaminated by Chernobyl's radiation. That left fish (almost always deep-fried in what was probably lard), vegetables (mostly cabbage and potatoes), and *khleb*, the good black bread that is Russia's staple.

After breakfast, the Americans came face-to-face with the crippling need that they would not be able to ignore. Vladimir Lipsky, chairman of the Byelorussian Children's Fund (BCF), sent a car to bring them to the BCF office. Lipsky was a charming, energetic man of small stature (perhaps five-feet-five) and enormous passion who described his organization as a "child of *perestroika*." While it received some funding from Moscow, it depended on local and international support from individuals and organizations of good will to do its work. As an "official" state agency, it was also free to relate directly to international agencies throughout the world with a minimum of restrictions.

"We are not too proud to ask for help," he told his guests through an interpreter in formal greetings at the BCF building on Communisteechiskaya Street in Minsk. They sat around a long conference table of blond wood. On Lipsky's desk, at the end of the narrow, sparsely furnished room, sat an old manual typewriter and a red telephone under a portrait of Lenin. The opposite wall was lined with blond wood bookcases with glass doors. It was nearly identical to every other office they would visit during the trip.

"The Children's Fund receives letters from the (Dead) Zone that cry out with a mother's pain: 'Our children have become so sad and so old. They often talk of death.'

"Another says, 'My granddaughter was operated on in the oncological hospital. She's an invalid when she's only four. What is her guilt? Why does she have to suffer?'

"You see, all our children are the children of Chernobyl." Lipsky captured the Americans immediately with his direct, guileless appeal delivered with warm smiles and broad, theatrical gestures.

"We are grateful that you have come to help," he continued. "The six district clinics in the contaminated regions desperately need diagnostic equipment to evaluate the health of children living there. They have excellent doctors but they don't have the tools, such as whole-body cesium counters and the like, to make the sophisticated examinations that are necessary."

Michael was scribbling notes on the blank pages of his journal. He was grateful for the pauses in conversation that the translation from Russian to English required.

"Then there's also the new Anti-Chernobyl Diagnostic Center where investigators are hoping to build a registry and data base on all aspects of the health of the children of Chernobyl," Lipsky said. "That's where we are concentrating most of the best equipment we have."

Paul smiled at the English translation of the name—the "Anti-Chernobyl Center." "Sounds pretty descriptive to me," he whispered to Michael.

The attractive marble and wood-paneled Anti-Chernobyl Center of the Radiological Medicine Research Institute was well-established and filled from morning to night with youngsters from the contaminated regions awaiting their twice-yearly examinations. This former clinic for the *apparatchik,* high-level Communist party officials and their families, was now a diagnostic center. The children were there for monitoring; they would be referred to other hospitals for treatment.

Medication was an ongoing need, Dr. Valeri Rzheutsky told the Citihope team. "We have enough doctors. What we need is medicine and new and more effective equipment. We need consultation with Western doctors and the opportunity to visit research centers in the West to learn the latest procedures and treatments. We also need hard currency to purchase the medicine our patients require."

The district clinics still needed virtually everything to carry out their diagnostic tasks. Lipsky suggested the answer lay in a mobile van that could carry the expensive and highly specialized equipment from district to district on a regular schedule.

Finally, Lipsky explained why the children must deal with daily life in a contaminated region. "It's not possible to find new homes for 2.2 million people who are exposed every day to various levels of radiation. We want to, but we cannot. The task is too great.

"So we must teach our children how to live with radiation, how to live happy lives without playing or even walking in the woods, without eating contaminated fruit and meat. They must supplement their meals with thyroid-enhancing vitamins and spend holidays in clean regions, away from the invisible radiation that surrounds them every hour of their lives."

The year before, more than one thousand children of Chernobyl spent the entire summer in camps in Western Europe. About one hundred children went to Cuba for a summer break and, according to Lipsky, two of them died of radiation sickness while they were there.

He mentioned that American actor Paul Newman had hosted eighty kids from Kiev in the Ukraine at his Hole-in-the-Wall Gang Camp for young cancer victims in Connecticut, and ten children with cancer from Minsk spent two weeks at an American Cancer Society/YMCA camp in Michigan—the group that had been profiled in the *New York Times* article Paul Moore had seen. Lipsky asked if Citihope could find more sponsors to provide summer homes and camps for thousands more children of Chernobyl.

Michael was intrigued by this enigmatic, fifty-year-old editor of children's magazines and author of many popular books who also had a reputation as an enthusiastic advocate for orphans in Byelorussia. At dinner that night, Michael found the opportunity to ask the question on his mind. "Why have you taken on this cause? The needs of the children of Chernobyl are overwhelming. How have you been affected personally?"

Turning serious, Lipsky answered with a story. "My own childhood was snatched away by the Nazis, who burned my village and my own home. I spent the war years of my youth living in the woods with my family, eating what we could forage and hiding from the fascists.

"When I grew up, I could identify with the needs of other children who had been deprived, too—first orphans, and then these children of Chernobyl who represent still another generation of Byelorussians who have had their childhood cut short by a tragedy they can neither understand nor control."

Before the Americans left the Children's Fund office that morning, Dr. Lydia Litvinovich, the ever-smiling, ever-fussing doctor on the staff, pressed some papers into Michael's hand.

"Please, would you consider this special case? She is a child of Chernobyl, a patient of mine. She needs more specialized treatment than can be found in the Soviet Union. Her name is Antonina Meshkova, a toddler of two and a half. She must have a liver transplant."

Dr. Lydia, a sturdy, rosy-cheeked woman with unruly, henna-colored hair, talked very quickly in imperfect English that Michael struggled to understand. "I've written a letter to Children's Hospital of Pittsburgh in the United States to see if they would do the transplant surgery. Please, would you contact the hospital and do what you can to encourage them to take Antonina?"

Among those papers was a letter from Vladimir Lipsky and Dr. Lydia that outlined the critical details of Antonina's short life. Born in January 1988, she was the only child of one of the "liquidators" at Chernobyl, the men who put out the initial fire after the explosion at the reactor, and his wife Svetlana. The father was now very ill and there would be no other children, the letter said.

Six weeks after her birth, Antonina was admitted to the Children's Surgical Center in Minsk with pronounced jaundice. The infant was diagnosed with biliary atresia. "This is a congenital pathology," Dr. Lydia wrote, "which we believe is due to a prolonged stay by the child's father in the area of the Chernobyl nuclear disaster."

"Infants like Antonina are expected to die rather quickly, but contrary to all prognoses, she has proven to be a fighter whose tiny body has refused to give in to the disease. According to Soviet surgeons," the letter said, "the child can only be saved by a liver transplantation."

Michael's head was spinning. He'd been in the country less than twenty-four hours and he was already on "system overload." Christmas presents and vitamins were one thing, but a liver transplant! He slipped the letter about Antonina into his travel bag. It was just one of several burdens he would take back to America.

❊ ❊ ❊

THE MORNING SESSION at the Children's Fund proved to be only a preamble to the most moving and, in the end, pivotal part of the three days the Americans would spend in Byelorussia—visiting the treatment centers and meeting the children.

At the Children's Hematological Center, Michael and Paul were captivated by Dr. Olga Aleinikova, a small, attractive woman of forty with short, reddish hair who no doubt kept some of her young patients alive by the sheer strength of her own will. Speaking near-perfect English with a British accent ("VITT-a-mins" rather than "VITE-a-mins"), Dr. Olga was animated and sometimes emotional in sharing the stories of the children she treated—mostly leukemia victims from throughout Byelorussia.

She led her visitors to each of the seventy-eight beds in the hospital and described the illness of each child. "Children with leukemia are very sick. They lose their appetite and they lose weight. They become pale and tire very easily. Often they develop diseases of the liver and spleen, and those organs become enlarged. They get fevers and skin disorders. And then most will die."

With her hands in the pockets of her white lab coat, Dr. Olga moved to the bed of a six-year-old girl who turned her head to meet the doctor's gaze. "When children are dying, it's terrible for doctors—for me, especially, when I know I cannot do enough without the medicines."

Dr. Olga caressed the girl's cheek with the top of her hand and the child nuzzled against it. "This is Ismaela. When we talk, she sometimes begins to cry, and I must be very strong. I tell her everything will be all right. Sometimes even I believe it."

Paul knelt next to Ismaela's cot and the child struggled to raise herself up. "Hello, Ismaela. My name is Paul and I have brought you a present from people in America who love you and hope you will get well soon." He dug into one of the bags Michael was lugging and handed her a small stuffed lamb. The child whispered a barely audible *"Spaseebuh,"* as she collapsed back on her pillow.

Dr. Olga told the mothers, who were sitting with vacant expressions next to the beds of their children, who these guests were, but she didn't apologize for the intrusion.

"Privacy," she said, "is something my children cannot afford. The world needs to know about my hospital and my children because I need the world's help."

The Americans were in a position to tell the story of the children of Chernobyl. So they would see everything—and she would show them.

It didn't take long to see it all. Her hospital had no working x-ray equipment, no lab of its own, very few of the potent leukemia drugs available in the West, and one overworked nurse for every fifteen patients. The building dated from 1931 but looked even older. From time to time, chunks of ceiling plaster and paint chips dropped onto the beds of the children.

Some reconstruction was going on at one end of the building. Dr. Olga was trying to make room for all the children who wanted to get into her hospital, but what she really wanted was a modern hematological center where she could practice the "modern protocols" she learned on recent trips to Germany and Switzerland.

But those weren't just learning trips; Dr. Olga was selling, too. She did a good enough job to convince doctors in Frankfurt to sponsor a telethon that raised 3.2 million Deutschmarks for her modern hospital dream. "They're keeping the money there until we have enough. Then they will build it. That's the way it has to be done," she said.

Until then, she was working her connections in the West to get "the most critical thirty percent of drugs that we are missing," as well as catheters, disposable syringes, IV "sticks" on wheels, and the other hospital basics that she used when she had them. Most of the time she had to do without.

Dr. Olga knew she had to work fast or more of her children would die. In 1986, the year of the catastrophe at Chernobyl, she had forty patients; in the cold, gray autumn of 1990, that number was nearly doubled and she was convinced she was many years from seeing the peak.

"We don't know what's coming, but we know it can only get worse. We are living an experiment. We were lucky in Minsk; the cloud (of radiation) didn't come here. Our air is clean, but our food is contaminated and we know it. We have no choice but to eat it or starve.

"We have to save these children! We are at least twenty years behind the West in what we can do for them, so we must hurry," she said. "Last spring, I went to Frankfurt. I was in the best training facilities in Europe for two and a half months. I was shocked that the children there were not cared for the same way as in the Soviet Union. Here we have all the books from all the Soviet doctors and Soviet academicians and we tried to do all they said.

"And then I found out that we were too much behind all those other countries." She spoke her words evenly, but she was agitated. She dropped her head into her hands as she sat at her desk. When she raised her head again, she clenched her fists in front of her face.

"It was a shock for me. Not your standard of living, but your hospitals and your treatments and what you could do for your children. This was my first time in those hospitals, but what about all the other Soviet doctors who go to those conferences and congresses in the West? Why didn't they tell me what they saw? I was angry!"

She ran the fingers of both hands through her hair as she recounted her outrage at knowing her children were dying because of officially condoned ignorance.

"In your country, eighty-five percent of most forms of leukemia can be cured. Here in the Soviet Union, only fifteen percent are cured. Why is this? Because we don't have the necessary drugs. See this bottle?" She held up a small vial of a clear liquid. "This is methotrexate. This costs us two hundred dollars. A child needs five of these each year to survive! If I had this medicine, eighty-five percent of my patients would live, just as in the West. I don't want my children to die!"

She had brought back from Germany samples of powerful leukemia drugs like methotrexate, along with the modern protocols with which to use them. She had coaxed a team of doctors and nurses from Switzerland to come back with her to train her staff. And she was determined to institute the new treatment techniques now—even if she couldn't do them all.

Dr. Olga was using highly toxic chemotherapy drugs without the tools or reagents to monitor them, but she couldn't afford to wait until all the pieces were in place.

"Lufthansa delivered those reagents six weeks ago in Moscow, but they haven't gotten to me here," she said. "I may never see them."

She told of a three-year-old leukemia victim who died of chicken pox while he waited for treatment.

Michael and Paul felt powerless in the face of such overwhelming need. Paul said, "We have to pray that people in America hear about your need and respond."

"I'm not sure I believe in prayer—or even in God—but I do tell the parents that when their child dies, they will see each other again in another world. But, of course, I don't know if it is so."

"It is so," Paul said quietly—and then he posed a question. "How many bottles of methotrexate would it take for you to believe in a God of miracles?"

"What do you mean?" she asked.

"A hundred bottles, two hundred bottles?"

"One thousand bottles!" she replied with a laugh and a toss of her head. "One thousand."

❅ ❅ ❅

NEXT STOP ON the American tour was the Oncological Hospital in Minsk, where they met one child of Chernobyl who was destined to move them in very special ways.

Natasha Ptushko was a weak and ashen-faced fourteen-year-old with pale blue eyes. She was lying in bed—alert, friendly, and eager for visitors. Eighteen months of chemotherapy for non-Hodgkin's lymphoma had robbed her of her white blonde hair and turned her teeth translucent gray. Still, Natasha's eyes were clear, shining windows into her soul, and her white face, framed by a rose-colored print scarf, was engaging and memorable.

As the three moved from bed to bed, Paul asked each child the same question: "If you could have a wish, what would it be?"

Each child responded in similar fashion: "I wish to be healthy again."

When PJ, in his role as cameraman, turned his videocamera to Natasha's bed, she struggled to pull a blanket over her thin body and hide the IV connection on her wrist that dripped saline solution into her vein.

"Please don't hide," PJ said to her. "I'm sorry for the intrusion, but we want to hear your story. It would help convince people in America to send medicine."

After Paul repeated his familiar question, Natasha paused a second and then turned her face to PJ and the camera. "Of course, I wish to be well, to be healthy. I want more than anything to be able to get up and be like all teenagers my age."

Looking directly at PJ, she said, "Just like you, I want to be healthy and I want to be loved."

Natasha's mother, nearly as thin as her daughter, had been huddled against the wall at the end of Natasha's bed. As Paul Sr. passed by, she reached out and touched his arm. Through her tears, she said, "We have been here a year and a half and still no cure. Help us."

"What is your name?"

"Alla, and my daughter is Natasha."

"Alla, many people will be praying for you and Natasha, beginning today."

"I want to see my daughter healthy and beautiful again, the way she was before, so beautiful."

"So do we, Alla. So do we."

Paul and Michael prepared to move on to the next room, but it was clear that PJ wasn't going with them. He handed the videocamera to Michael.

"I know I have a job to do, that I'm supposed to be the cameraman of this team. But I have to stay here. I can't explain it. I just have to stay here."

Paul started to object, to play the father to a rebellious son, but he stopped. He put his arm around PJ briefly and then turned to go. "We'll come back for you when we're ready to leave."

With Alla watching protectively from the end of the bed, PJ pulled a small chair up to Natasha and opened his backpack. An interpreter crouched near the bed. PJ took out a cassette tape player and a tape of popular American songs he had edited himself.

"This tape has one of my favorite songs." PJ slipped the cassette into the player and turned it on. "It's called 'Love.' The words are simple: 'Love is patient; love is kind. No eyes of envy, true love is blind.' "

Natasha listened as the words were translated and then listened to the music. The song and the conversation seemed to energize her.

"Say the words to me in English, slowly. I want to learn them in English. I will memorize the words in English and sing them to everyone. I used to sing all the time, before I was sick. Now I will sing again—and in English!"

Over and over, PJ repeated the phrases of the song, and Natasha began to open up to this tall American teenager. They talked about their families and their schools.

PJ asked her if she remembered the day of the explosion at Chernobyl.

"Yes, I was nine years old. I was at home. There was a radio announcement that there might have been some accident at the reactor. No one said that it might be dangerous. I remember thinking it might be. But we were busy getting ready for May Day and the parade. Now we know why it was so hot that spring, why even the rain was warm when it should have been cool. Now we know how dreadful it really was."

Finally, she began talking about her illness. "I've started to lose hope of surviving. I don't know what can happen. It seems to me that I might die soon."

That may have been the first time Natasha had confronted her own death. Soviet doctors generally did not tell their young patients that they had cancer, and they encouraged parents to avoid the subject as well. "We think it's better not to frighten them with thoughts about death," one doctor explained.

But, of course, Natasha knew, even if everyone around her kept the secret, and she chose to share her knowledge with this caring young man from New York.

The visit wasn't long—Paul and Michael poked their heads into the room in half an hour. But the children of Chernobyl now had a face for the Americans—the face of a "withering white rose," Michael called it. That face belonged to Natasha, and the image of her wan smile and hope-filled eyes was imbedded in their consciousness.

The image was even more enduring for PJ; for the first time in his young life, he was falling in love.

❋ ❋ ❋

GLASNOST, the new openness in the Soviet Union, and the economic restructuring of *perestroika,* meant different freedoms for different people. For Viktor Tushikov, a young interpreter working with the Citihope delegation, it meant unusual freedoms for the press. One morning at breakfast at the Belarus Hotel, he rushed up to Michael with an erotic newspaper, a poorly reproduced *gazyeta* with a nude female model on the front page. Viktor glowed with excitement.

"Now this, *this* is *perestroika!"* he crowed.

Michael handed it back with a shrug and a wry smile. But he did marvel that the same openness that permitted nudes on newspapers in a nation as prudish as it was atheistic also meant that Bibles from the West no longer had to be smuggled in past border guards. The just-signed legislation on freedom of conscience in the Soviet Union now also guaranteed the freedom to worship, proselytize, receive the sacraments, teach Sunday school—and marry in a church, if one wished.

When Viktor told Michael the news of his recent engagement, Michael asked him if he would be married in a church. "I don't think so," he replied. "I'd like to, but I'm not a believer and I don't think it's right to get married in a church just because it's the popular thing to do."

WHAT ENCOURAGED Paul and Michael the most on their trip was seeing firsthand the renewal of religious interest in the Soviet Union and the renaissance of the Russian Orthodox Church. As "priests from America" they had been invited to spend a Saturday evening with the metropolitan of the Orthodox Church in Byelorussia.

Metropolitan Philaret was born in Moscow in 1935. A permanent member of the Holy Synod of the Russian Orthodox Church, he had held the post of patriarchal exarch to Central and Western Europe before coming to Minsk. A popular and revered figure, he also served as an elected people's deputy in the Soviet Congress in Moscow.

The three Americans were thrilled with the invitation. They all felt they were being treated with honor beyond their status and accorded privileges usually reserved for dignitaries. They were unaware at the time that such a reception was "part of the tour" for Western visitors who had expressed genuine interest in providing help for the children of Chernobyl.

The three men were guided through the opulent private and public rooms and chapels of the headquarters of the Russian Orthodox Church in Minsk. Philaret explained the history and significance of different altar pieces, including an incredibly beautiful gold iconostasis, which the Orthodox considered to be the window to heaven. It also served as the barrier beyond which only an Orthodox priest could step.

The interpreter, Viktor, was having a difficult time with the religious terms Philaret was using. When he interpreted *svyatoi* as "god" rather than "saint," Philaret angrily chastised him. Apparently, Philaret understood more English than he let on, and he obviously would not put up with less than accurate interpretation. He rebuked Viktor for his ignorance of things holy, and their relationship remained strained all evening.

The guests were ushered into a spacious, classical dining room where a large table with twelve place settings held the makings of an elegant feast, including black and red caviar. The room was dominated by a huge painting of the Last Supper, and Michael couldn't help notice the obvious symbolism as Philaret took his seat in the middle and Paul Moore, apparently following protocol, was seated directly opposite him. The other guests— Vladimir Lipsky, Alexander Trukhan (second in command at the Byelorussian Children's Fund), Alexei Zhylski (Byelorussian government supervisor of religious activities), Viktor, members of Philaret's personal staff, Michael, and Paul Jr.—all found their places.

Dinner conversation centered around the relationship between the church and the state and Philaret's fear that the democratic movement might take the country too far. "There is a dark side to freedom. *Perestroika* can be a dangerous thing," he said. "It wouldn't be a good thing, for example, for the Church of Satan to get registered."

After the first round of speeches and toasts, which were becoming familiar to the visitors, Paul asked, "We have heard rumors that the Orthodox Church has grown dramatically during the last five years under Gorbachev and *glasnost*. Has it really been that dramatic?"

Philaret instantly became animated. "We have new churches in every district of the republic. Two years ago we had 356 churches in Byelorussia," he said. "Today we have 525. Churches that had been confiscated by the state are being returned to the custody of the church every day. New theological schools are needed to train new priests. Most of our churches are full on Sundays.

"Best and most important of all, we have been able to reopen the Sunday school at the cathedral in Minsk. Until this very month, it was illegal for us to provide any kind of religious training for our children."

Viktor, eager to get it right, interpreted without correction from Philaret.

Zhylski, the Communist government representative, rose to initiate the second round of toasts. "A few years ago, my job was to watch this man (he raised his glass to the metropolitan) and to make sure he wasn't doing anything wrong. Now my job is to help him assist all the churches in their religious activities. Now that religion is legitimate, it's a whole new day in the Soviet Union."

He didn't comment on the obvious contradiction that the right to practice religion had always been included in the Soviet constitution, but patently ignored. All the oppression that the church had weathered during seventy years of Soviet control had been unconstitutional from the outset.

The conversation finally turned to the serious nature of the visit: how the Americans could help the children of Chernobyl and the Byelorussian Children's Fund.

Lipsky described the need poignantly and Paul responded expansively with a pledge of compassionate response. Then His Eminence looked at Paul with great intensity and urgency. With his bulging eyes staring out over his unruly, gray-black beard, he declared: "We are one!"—meaning the church, the BCF, and those who take up the cause of the children of Chernobyl.

Paul answered with equal intensity: "We are one!"

Through this exchange, Michael was squirming in his chair. He was a bit intimidated by the personal power and magnetism of the metropolitan. He admired Paul for rising to the challenge of relating to Philaret as an equal, but he was profoundly embarrassed that the metropolitan obviously saw them as something more than ordinary American clergymen with decidedly limited resources on a journey of personal discovery. He was afraid they were promising more than they could deliver, and he was equally worried that Paul didn't seem a bit concerned.

Michael noticed that Philaret could be jovial and childlike one moment and then turn dark, intense, and overpowering in an instant. Like Billy Graham, the Dalai Lama, or Mother Teresa, the metropolitan was larger than life, and this proved disconcerting.

While Michael and Paul were inspired by the awesome presence of Philaret and the power of his position, Paul Jr. sat silently throughout the meal. Later, as they rode back to the hotel, Paul asked him what was up.

"I can't understand why Philaret would surround himself with all that gold, all that obvious wealth, when he knows that children are dying. And I can't understand why you and Michael were sucking it all up," PJ said. "We had just come from the cancer center. Dining in splendor after being with dying children is just too much.

"Why couldn't the metropolitan sell an icon or two to help Natasha?" he asked. And why was everything so different from what he had expected? Although he didn't express the thought in so many words, PJ was feeling confused and unsure of all the things he had been so absolutely sure of when he left New York. Nothing in the Soviet Union was the way it was supposed to be. Michael resolved that he would try to help PJ talk about his unsettled feelings once they had begun their long plane flight home.

The trip had in fact left all of them with far more questions than answers, but they headed back to New York with a consensus decision nonetheless: the children of Chernobyl would be the focus of Citihope's international outreach effort.

Checking out of the Hotel Belarus under a leaden autumn sky, they were no longer hauling the ten bags of gifts and medicine. That cargo had been delivered to the children they had met along the way. Instead, the three carried a far heavier burden—the hopes of Natasha, Antonina, Ismaela, and the other children they could neither forget nor ignore. They had signed on, but it would still be some time before they discovered the scope of the commitment they had made.

3

"DON'T SEND ME
ANY VITAMINS"

TWELVE UNITED CHURCH OF CHRIST peace ex-
change delegates arrived in Minsk on the overnight train from
Moscow on a bright, cold November morning. About a month
after the visit of the Citihope team, they came on their own mission of citizen-diplomacy and peacemaking, unaware that they
were walking in Citihope's footsteps.

Their hosts, a representative of the Minsk Peace Committee
and Tanya, a high school English teacher, greeted them with red
carnations and warm smiles as they climbed off the train, even
after one member of the group mistook them for flower vendors
and waved them away.

Attractive and smartly dressed women in their early forties,
they laughed off the mistake and persisted in connecting with
the UCC delegation, the fifth in a twelve-year-old series of ex-
changes between the Northern California conference of the
church and the Soviet Peace Committee.

The delegation was led by two ordained UCC pastors, Carolyn Roberts from Menlo Park, California, and Nell Carlson of
Seattle, Washington.

Among the other clergy and lay delegates was Michelle
Carter, a San Mateo, California, newspaper editor who had followed the Chernobyl story since the explosion and had been a
student of the Russian language, and all things Russian, since
high school. Michelle was in her mid-forties, and with her jour-

nalistic training and commanding presence she had a knack for getting what she wanted.

Close friends also knew her as Micki, someone who could be counted on for energy and an upbeat attitude. With a camera, notebook, and tape recorder in hand, she had both a professional and personal interest in traveling to the region. This was her second trip to the USSR. She had none of the conflicting and anxious feelings that had troubled the men of Citihope upon crossing the Soviet frontier. Instead she was excited about the chance to travel through the lands of Pushkin, Tergenev, and Yevtushenko. She prided herself on her ability to be an objective observer of culture and human nature, but on this trip it would sometimes take all the strength she could muster to control the floodgates of her emotions.

Carolyn, Nell, and Michelle were veterans of a 1988 delegation that had taken part in the celebration of the millennium of the Russian Orthodox Church—one thousand years since its founding and the birth of the Russian empire.

That trip had taken them to Tallinn, the capital of Estonia, where they had witnessed the wobbly, first steps on that Baltic republic's march to freedom; Peter the Great's showplace capital, Leningrad (which has now reclaimed its czarist name, St. Petersburg); Stalingrad (now Volgograd), the site of the pivotal street-by-street battle of World War II; and Moscow. In 1990, the delegation had made stops in Moscow and the medieval "Golden Ring" cities of Vladimir and Suzdal before heading for Byelorussia and Minsk.

In 1990 Byelorussia had been part of the Russian empire for a mere two hundred years—since the three "partitions of Poland" in the late eighteenth century shoved the republic into the Russian sphere of influence. However, the tumult of the Russian revolutions of February and October 1917 allowed Byelorussia to declare its independence in 1918, which held until the Red Army overran the rebellious state and forced it into the new Soviet Union as the Byelorussian Soviet Socialist Republic.

Byelorussia enjoyed another brief period of self-rule at the end of the Great Patriotic War in 1945, shortly before the rout of the German occupation forces by the Red Army. It was this brief sovereignty that allowed it to be admitted to the fledgling United Nations as a state separate from the USSR. It continues to hold that membership today.

The last-minute addition of Minsk to their itinerary would have a profound impact on the lives of some members of the United Church of Christ group. They shuddered as Tanya, the English teacher, spoke in front of the bronze blacksmith, the statue of the lone survivor that stood guard at the entrance to the village of Khatyn. "He is holding the charred body of his son, whom he carried out of the barn where more than one hundred Khatyn villagers were incinerated by the fascists." She spoke so softly and reverently that the Americans had to lean close to hear her.

Huddled against a biting November wind, they walked silently past the individual plots that marked the houses of the village, gates standing open to welcome the visitors. Mournful, atonal bells sounded every few seconds in a dirge to the unnamed dead.

"This memorial is much more than a monument to a single village destroyed by the occupying forces," Tanya continued. "It commemorates all the Byelorussian villages, one hundred and eighty-six of them, that were burned to ashes as the Germans sought to eliminate an entire people."

Later, back in Minsk, Tanya translated the words of V. I. Lenin on the wall of the Green Cottage, across the street from the BCF, where in 1898 the first Democratic Socialist Workers Congress was held in great secrecy. On a path in a downtown park, she recited the impassioned verses of the Byelorussian national poet Yanka Kapala. At the Byelorussian National Theater, they leaned close as she whispered bits of the plot of a Tchaikovsky opera, "The Queen of Spades."

"I planned to take you to Lenin Square to see the Lenin monument, just as I take all first-time visitors to Minsk," Tanya announced as the Intourist bus parked across the broad avenue from the square. "However, this time the square is closed. After the demonstrations there on November 7, the Revolution Day holiday, party officials are not allowing anyone in the square."

Revolution Day is the anniversary of the beginning of the Russian Revolution in 1917 and, as it turned out, 1990 had witnessed the last military observance of that day in the USSR—one of a long list of "lasts" and "firsts" in the Soviet Union in 1990 and 1991.

"Party officials had expected trouble in Red Square in Moscow, where tens of thousands of people turned out to protest the serious economic conditions of our country," Tanya continued. "But there was no violence in Moscow. Lots of tension and confrontation, but no violence. 'Vremya,' our nightly news program, reported only one incident of violence in the entire country—in Minsk, in our usually very quiet republic of Byelorussia.

"It really wasn't very dramatic. Members of the Byelorussian Democratic Front wanted to put flowers at the Lenin Monument, when the army moved to stop them. No one was killed, but I think the world now has a different opinion of our republic. Perhaps we aren't so boring and uninteresting anymore."

The delegates also stopped at a beautiful red brick church just off Lenin Square which was under reconstruction as a Catholic cathedral. The unusual building was constructed in 1915.

"Then came World War I and then the revolution," Tanya said. "It was never sanctified. Now it has just been returned to the Catholic Church and they are holding services in a small chapel while it's being rebuilt."

Inside, some men and women of the congregation led the group on an impromptu tour of the cathedral, explaining where things would be and how they would look when the reconstruction was completed. Several of the delegates dropped handfuls of rubles into the collection bowl to help with the project.

"There's nothing to buy with the rubles," they joked among themselves. "Why not rebuild a cathedral?"

Finally that day, they too met Dr. Olga, who charmed them and challenged them just as she had the three from Citihope and, surely, every other group of Westerners on official visits to Minsk. The local Peace Committee had a mission to tell the story of the children of Chernobyl, and Dr. Olga could convey that story and the long list of unmet needs with a directness and passion that appealed to visiting Americans.

Her command of English was certainly significant, but she also seemed to understand and even anticipate the American need to ask endless, sometimes pointed questions in an effort to comprehend the scope of the tragedy. She was neither offended nor bored, and she made each guest feel as though his or her efforts for her children would alone determine whether they lived or died.

Dr. Olga was specific about the things she did and did not need and precise with her shipping instructions. "If you know where to get catheters, you must send them to me here at this address. If you send them to the Ministry of Health in Moscow, there's no guarantee I will get anything. . . .

"And mark everything for the children of Chernobyl. Then it will get to me."

One mother was waiting in the hall outside Dr. Olga's office while she took her guests around. When Dr. Olga passed by, the mother caught the sleeve of her white jacket, sobbing and imploring her. Later the doctor, who has two teenage children of her own, explained that she would take the woman's child, even though she lived in another republic, the Ukraine.

"When a mother cries, I cannot tell her 'no.' Here I can offer mothers hope when they have no hope. . . . That's why I must build here a new, modern center so I can take all the children of Chernobyl."

She hadn't missed any other angles, either. Although she told her visitors she was not a believer, she had observed church volunteers working with the children in the hospitals she visited in Germany and Switzerland. So, when she returned to Minsk, she charmed Russian Orthodox nuns and priests into making regular visits to her hospital, even though hospital visitation wasn't common practice for clergy in the Soviet Union.

"Did you see the children's artwork on the walls in the playroom?" Dr. Olga asked. "One of the nuns comes every few days now and works with the children. I guess you would call it a sort of art therapy—although I've never heard of anything like that anywhere else in the Soviet Union."

There was nothing artificial or contrived about Olga; no one doubted for a second that she cared deeply about each child at the Hematological Center, that each child's death was a personal wound.

The example of her strength made it possible for some of the delegates, most of them parents of young children themselves, to hold back their tears as they walked into the rooms. While some of the children were giggling and squealing, behaving like normal youngsters, others were too weak to lift their heads.

Eight-year-old Elena whimpered as Dr. Olga noticed the redness and swelling around the IV connection on her hand and called for a nurse to insert a new one.

Six-year-old Volya screamed and his mother sobbed as she attempted to inject more medication into a shunt in her son's chest. "Why isn't a nurse or a doctor doing that?" one of the delegates asked.

"In Western hospitals they would," Dr. Olga explained. "But here our hospitals are so short-staffed that mothers have taken over all but the most sophisticated tasks. Also, it's part of our culture for mothers to continue to be responsible for their children—even when they are hospitalized. They don't surrender their responsibility to anyone, not even to doctors and nurses."

The Americans sang "Happy Birthday" to six-year-old Sergei as much to lift their own spirits as his.

A *babushka* (literally "grandmother") hovered over her beloved granddaughter, Nadia, while Dr. Olga explained, "She's just sixteen and she had been in remission (from leukemia) for more than a year when she was rushed back here. She had a massive intrauterine infection and she had to have a hysterectomy a week ago. But," she paused as she brushed back Nadia's thick blonde hair, "the surgery revealed new, more serious tumors."

In the hall outside Nadia's tiny, private room, Dr. Olga whispered, "The first time they come to us, we have a real chance for success. But the second time they come (after remission), the chances are much less."

When the delegates returned to the dimly lit foyer of the Hotel Belarus that night, they gathered in one of the cold rooms for a brief time of worship to provide closure to a heartbreaking but also motivating day. The Rev. Matt Broadbent of Santa Cruz, California, lent his magnificent baritone voice to lead the group in a chant based on Micah 6:8: "What does the Lord require of you? To seek justice, love mercy and walk humbly with your God."

Then they all dispersed to endure the night in their own freezing rooms.

❊ ❊ ❊

"I CAN'T BELIEVE it's this cold *inside*," Michelle said to roommate Carolyn Roberts as she wrapped the blanket from the bed around her. She still hadn't taken off her heavy coat, muffler, and hat.

"There's a big gap here under the window," Carolyn said as she knelt, half-hidden by the thin drapes. "I think I can stuff some newspapers in here to at least block the worst of it."

"I'm going to ask the *dzhernaya* for some more blankets," Michelle said as she unwrapped herself and headed down the hall to the floor attendant. When she reappeared in the room, she had two more blankets in hand. "She said the whole city is heated by a central system and it's broken. No one in Minsk has any heat tonight. I'm putting on as many layers as I can and getting into bed."

Carolyn mumbled a reply from under the blankets in her narrow bed, which was placed at a right angle to Michelle's. Her gloved hand crept up the wall to the toggle switch and the lights went out.

Michelle couldn't shake the images of the pale, swollen-faced children she had hugged that day and the despairing, nearly numb expressions of their mothers. She was frustrated that her command of Russian had been insufficient to express adequately the compassion she felt.

As she reviewed their faces in her mind, memories of her own pain surged to the surface—memories of the long hours she and her husband had spent at the crib of their daughter Robyn during the many hospitalizations she had endured. Robyn's condition, a congenital urinary tract impairment, was not life-threatening, and she had since grown into a healthy young woman. But the pain a parent endures, listening to the cries of her child whom she cannot comfort, was reborn in her that autumn night in her cold room.

She felt compelled, perhaps even directed, to do something for these children, to act on their behalf. Her experience with Robyn's illness had prompted her to become an advocate once before. Upset that she had to fight for the right to stay at Robyn's bedside at all hours in the hospital, she investigated and wrote a series of newspaper articles for the *San Mateo Times* that eventually led to a change in hospital policy. Now, parents are free to be with their hospitalized children at all times in that northern California hospital. Putting the public spotlight on an ill-conceived hospital policy had resulted in positive change.

Of course, she intended to write as extensively as possible about the plight of the children of Chernobyl, but perhaps there was more she could do. With leather gloves on her hands, which were stiff from the cold, and with her purse-sized flashlight for illumination, Michelle began to jot down in her reporter's note-book the outlines of a project to raise funds to buy methotrexate and locate a source of disposable syringes, catheters, and IV "sticks" on wheels that would allow those children getting chemo-therapy to get out of their beds and move around. San Mateo County, where she lived and worked, was home to dozens of bio-technical and biochemical companies, and she brainstormed ideas for reaching them with an appeal for the children of Cher-nobyl.

She also made a note: "No vitamins." Dr. Olga had been very clear. "Don't send me any VITT-a-mins. That's all anyone ever sends. No, send me methotrexate. Put my name on it. Send it di-rectly to me here."

Despite their obvious value for a childhood population at risk for immune deficiencies, Dr. Olga had set her priorities. There would be no vitamins on Michelle's list. Only methotrex-ate and the reagents to maximize its efficiency.

❊ ❊ ❊

BEFORE DROPPING OFF to sleep, Michelle also made some notes about another child of Chernobyl named Zhenia whom she had met at an earlier stop on the trip.

In Vladimir, the ancient city of feudal princes and Mongol raids in the historic Golden Ring around Moscow, the delegation had visited the local Pioneer Palace—an institution that provides day care, activities, and after-school sports all in one building. The Americans had spent some time with a group of teenage girls in the International Relations Club and were saying good-bye to the club's director, Tatiana Leskina, when she was asked if she planned to continue her work at the Pioneer Palace.

"Oh, of course. My husband was at Chernobyl. I don't know how long he'll be able to work," she replied quite matter-of-factly in her school-girl English. "And I have a son."

Pressed for details, Tatiana said her husband, Sasha, was in the army, one of a company of four hundred young men from Vladimir assigned to maintenance duties at the graphite-moder-ated nuclear reactor in the Ukraine when the catastrophe oc-curred. He was nineteen at the time. "Just two years older than David," Michelle remembered thinking at the time. Her own healthy and athletic son was a college freshman.

Sasha volunteered to stay on for four months and help with the cleanup and evacuation—four uninterrupted months in the area of the greatest radiation density.

When Michelle asked Tatiana if she would be willing to be interviewed for a newspaper story about the effects of Chernobyl on their lives, she agreed. A meeting was arranged for the next day at one of the scheduled stops for the delegation—a children's hospital in Vladimir. When the red Intourist bus carrying the Americans pulled up in front of the hospital the next morning, there were Tatiana and her son, bundled against the aching November cold, waiting and waving.

Tatiana and Sasha were not yet married in the spring of 1986, but when they did marry some months later, Tatiana remembered being uncertain about how her husband's exposure to radiation at Chernobyl might affect their lives. "No one knew for sure if there would be a problem but, of course, we knew there was a chance (of future illnesses).

"They were volunteers; they didn't have to stay," she said in between attempts to quiet three-year-old Zhenia, who was pulling kopecks from her purse and bouncing them on the floor.

She ran her hand over his cropped blond hair and added, "At the time, (the survivors) felt a fear that it might influence their children, their future children.

"I wasn't very worried when I got pregnant because Sasha wasn't sick but, nonetheless, I was relieved when Zhenia was born healthy. Now Sasha has started having very bad headaches and the doctors are studying his blood very carefully. So now I am worried—about my husband and about what will happen to my son and me."

Tatiana was quick to point out that the government hadn't turned its back on the Union of Chernobyl, an organization formed to simplify efforts to find and deliver medical attention and other forms of assistance to survivors.

"They gave us extra vacation days, and in July, the minister of defense ruled that the survivors of Chernobyl would have special privileges—they could ride the buses and Metros free and get apartments," she said. "However, it's difficult for us to use these privileges in Vladimir. No one has offered us an apartment, so we still live with my parents. . . . Only promises so far."

Tatiana added that Great Britain had sent special radiation-detection devices to a medical center in Kiev in the Ukraine to monitor the health of the Chernobyl victims.

"The survivors from Vladimir go there once a year for evaluation and, so far, only one soldier from Vladimir has actually become sick from radiation sickness," she said. "Of course, that doesn't count Sasha. They haven't told us that he is sick because of Chernobyl, but the two of us, we know it."

Before the catastrophe at Chernobyl, Tatiana told Michelle, she "had no idea that such a thing could happen. We thought it was all very safe. We didn't realize the truth until later."

However, the interpreter who was helping with the interview—a twenty-two-year-old institute-trained Muscovite—said he knew nuclear energy could be very dangerous long before Chernobyl. "I remember the concern about the accident at Three Mile Island in the United States. Now, after Chernobyl, there is a very strong movement against the presence of these reactors in the Soviet Union," he said. But not in Vladimir, where ordinary, industrial toxins have killed all the fish in the placid, ice-choked River Kliazma, a river where freshwater pearls were once harvested for the exquisite icon embroidery on display in the local museum.

"Here we have only coal energy," Tatiana said as she pulled mittens on the pudgy hands of her son and zipped his snowsuit up under his chubby chin. "There are no protests here at all."

STILL HUDDLED under the covers with her flashlight, Michelle closed her notebook and rummaged for a news article she had brought with her, the story of another man, a fifty-year-old nuclear physicist, who knew the real story of the cleanup of Chernobyl and the conditions that now were threatening the health of Tatiana's young husband. She wanted to read it to Carolyn, but in the moonlight only her curly, dark hair was visible above the covers and her breathing was even and deep. So Michelle reread the article herself, with deepened concern now that she knew a real family whose health had melted down as surely as Chernobyl's reactor No. 4.

Vladimir Chernousenko had found it necessary to leave the Soviet Union and seek asylum in Europe—France, Germany, and eventually Great Britain—because scientists in the USSR no longer wanted to hear his story. On October 14, 1991, an article about him appeared in the *New York Times*. The writer, Peter Matthiessen, had heard Chernousenko speak at an environmental conference in Mexico.

The scientists, environmentalists, and writers at that conference had paid rapt attention to this man's story of the catastrophe of Chernobyl, because he had lived it. Dr. Chernousenko had been the scientific supervisor of the emergency team that had been ordered into the area five days after the explosion to oversee the liquidation of the fire and cleanup of radioactive materials. He was also designated director of the total-exclusion, thirty-kilometer zone around Chernobyl.

When he arrived, the fire was still burning and the "liquidators"—mostly conscripted army laborers, like Tatiana's husband—were still working.

Matthiessen wrote:

"Dr. Chernousenko's orders were to 'liquidate the consequences' of the accident—he enunciates this bureaucratic euphemism with irony and despair, because the consequences will remain unliquidated for millennia—and to reactivate the other three reactors (which had been shut down after the explosion) as soon as possible, at any cost. . . .

"From the start, it was known that the amount of energy produced by the restarted reactors would be insignificant, that there was no valid reason to send people back into Chernobyl and every reason to evacuate the region. Dr. Chernousenko and his team personally warned Mikhail Gorbachev that a premature cleanup would drastically increase the human damage.

"Even so, the government decided that the cleanup should not await the arrival of modern technology and machines, but should start at once with manpower and shovels. . . . "

Dr. Chernousenko said that terrified, untrained army reservists and local coal miners were sent out onto the roof of the still-burning reactor to scoop smoldering graphite waste down into the reactor building before it was sealed. They were allowed on the roof only a minute at a time.

To illustrate the madness of this action, the *Times* reported, the physicist ran a film that "showed bulky figures running out on the roof. There was only time for two small, frantic scoops with narrow, old-fashioned shovels before they fled—'10th century technology being used to fight a 20th-century catastrophe,' " Dr. Chernousenko said. Clutching official certificates of honor, these young men were immediately removed from the

Dead Zone. Even so, without exception, those sent out on the roof are dead."

Dr. Chernousenko himself was the only survivor of the one hundred or more who actually put out the fire at Chernobyl—and doctors had given him only eighteen months to two years to live. Officially, the Soviet Union attributes only one hundred and forty deaths to the accident at Chernobyl, all of these occuring in the explosion or as a result of the liquidation. The physicist maintained that the real toll was more like five to seven thousand, and "many thousands more throughout the southern Soviet Union will die of radiation poisoning or related cancers, especially in Byelorussia," which received seventy percent of the fallout—far more than the Ukraine.

Dr. Chernousenko began to spread the word that Chernobyl represented the edge of "an apocalypse." He disputed the casual response of the International Atomic Energy Agency as an "attempt to assuage the public's increasing uneasiness about the safety and practicality of nuclear power."

He believed that the Chernobyl catastrophe had eliminated the last hope for the practical use of nuclear power, which he considered "the most dangerous threat faced by the world's environment."

They began to call him "the madman of Chernobyl," and soon Dr. Chernousenko was no longer welcome in the Soviet Union. He agreed that perhaps he was mad, but no matter. He intended to spend what was left of his life telling the story of Chernobyl.

No one else could tell it; he was the only one left.

Michelle folded the clipping, tucked it away along with her notebook, and turned off the flashlight. She drifted off to sleep, despite the cold.

❊ ❊ ❊

THE NEXT MORNING, the delegates piled onto a bus in a fierce Byelorussian wind for the short trip past Dr. Olga's hospital to meet the metropolitan of the Russian Orthodox Church, much as the Citihope group had done a few weeks earlier. On the bus, Michelle tried to analyze what it was about Dr. Olga that had motivated her and the others. They had met lots of people with many needs, but nothing had captured their attention the way this one doctor had.

Perhaps it was a natural gift, a sense for marketing, but Dr. Olga had presented the need and then sent the Americans off knowing precisely how they could respond. That kind of insight into the American psyche—and the ability to tell her story without the awkward intervention of an interpreter—was allowing her to succeed where others could not.

Michelle leaned over the bus seat to talk to Carolyn. "You know, Dr. Olga may have some sixth sense that allows her to understand Americans, but I'm convinced most Soviets have a hard time grasping our motives.

"Were you part of that long bus conversation some of us had with Oleg on the last trip?" During the 1988 visit, Michelle had commented casually to Oleg Tumin—a young, English-speaking Muscovite who teaches Arabic at Peoples Friendship University and served as a Soviet Peace Committee interpreter for visiting Americans—that Soviets don't seem to have a handle on what Americans are about.

"Yes, I remember that Oleg said part of the problem was that few Americans have the skills to get around the Soviet Union by themselves," Carolyn said. "In France or Italy, tourists may not understand the language, but at least they can read the Metro map and the names of the stops. But in Moscow, unless they have studied the Cyrillic alphabet, they can't even do that."

"So it's no surprise," Michelle said, "that the image Soviet citizens have of Americans is one of well-dressed tourists who move about the country insulated from the daily Soviet experience inside one of these red Intourist buses."

"Or the image they get from the 'things' America sends to the Soviet Union," Carolyn added.

Since Oleg had had the opportunity to get to know Americans on a personal basis and even to visit their homes in the United States, he had become increasingly concerned about the image of America among educated Soviets.

"He believes America is perceived in the USSR as a shallow and trivial country because the American goods in their homes are generally in bad taste—detective novels instead of great literature, T-shirts of poor quality, pornographic films," Michelle said.

"He knows that quality products and ideas, as well as great films, exist in our country, but he said they don't show up here.

To his friends, America has no substance because what they see of things American has no substance.

"But there's more to it than cheap American goods," Carolyn said. "It has a lot to do with the Russian 'soul,' I think. Russians love to debate and argue philosophy and politics. We both know it's not unusual for a Moscow dinner party in a crowded, two-room apartment to go on all night as the guests drink, smoke, and debate until dawn."

"One of Oleg's favorite childhood memories," Michelle recalled, "was going with his grandfather to the public baths where he would listen to truck drivers and bricklayers pass the time arguing over politics and poetry. He said Americans don't seem to share this passion for wrestling with great philosophical issues. He thinks this contributes to their shallow image among Russians.

"I wonder if he still feels that way? I wonder if they all do?"

When the bus stopped at Philaret's residence, the delegation was herded through the opulent private chapels just as the Citihope team had been. Those same concerns that had disturbed PJ bothered these Americans, whose Puritan roots took them back to the plain—some would say severe—religion of the early colonial era. Church money was best spent on direct aid for those in need; a gilded iconostasis was not their idea of service to God.

The notion of the church standing apart, without a direct response to obvious human need, proved to be a difficult concept for this religious delegation to grasp. In the United States, most churches are at the forefront of one outreach effort or another. Americans are open to appeals for money or volunteer service, and their gifts represent a greater percentage of their average income than anywhere else in the industrialized world. Identify a need, and somewhere an American or an American church would raise money for it. It's a uniquely American response, and it is difficult for Americans to understand that this response isn't universal.

Charity in the Soviet Union had been just another gear in the central planning network. "Donations" to political causes such as peace funds were deducted from paychecks or collected at the workplace; charity fund-raising was unnecessary since, the party maintained, there were no health or social needs left unmet by the state. Kopecks left in the collection plate at Russian Orthodox services were meant to sustain the priest and the

church itself; church-sponsored charity was prohibited by law—
again because the state would provide.

Even in disasters like the horrific Armenian earthquake of
December 1988, when people around the world sent money,
clothes, medicine, food, and blankets, Muscovites could not
recall any similar relief drives in the Soviet capital. The state
would provide, and most Soviets assumed that the massive airlift
of medicine and supplies—even the teams of dogs who sniffed
for victims in the rubble—were the gifts of other governments,
not of individuals moved to a compassionate response.

A year later, Andrei Sidorin, the Soviet reporter who was
chief of the Tass news service bureau in San Francisco, tried to
explain the same phenomenon to the *San Francisco Chronicle*. He
said he went to special effort to write about the incredible individual volunteer efforts that arose in northern California after
the Loma Prieta Earthquake in 1989 because it was a difficult
concept for his countrymen to understand.

"The significance of the human factor . . . is hard to exag-
gerate. . . . In the Soviet Union, probably because of the increasing hardships, people are more intolerant of each other, more
angry and excited. They have very little sympathy for one another."

So it should have surprised no one that the metropolitan of
Minsk spent the church's money on golden icons instead of or-
ganizing fund drives for the children of Chernobyl. The Soviet
system had sanitized the notion of charity—and even the idea of
unmet need—from the culture. In the fall of 1990, most Soviets
(and Americans) knew that the state could no longer provide for
a whole panorama of needs, including the children of Chernobyl. What surprised the Americans was that the Soviets no longer
knew how to respond as individuals.

The session with Philaret was their last stop in Minsk.
Michelle and Carolyn packed in silence that afternoon, each lost
in her own thoughts. After an early dinner, they left for the train
station.

❋ ❋ ❋

AS THE NIGHT train pulled out of Minsk, taking them to
Moscow and closer to their flight home, Michelle was wide
awake. She listened to the even breathing of her cabinmates,
then pulled back the curtains from the window and propped

herself up on pillows to watch the countryside by moonlight. She couldn't see the bright colors of the elaborate gingerbread trim around the windows of the cottages, so she filled in those details from memory. She waved to the woman sitting in the open window of a by-station, wrapped in heavy sweaters and a scarf, waiting for the train to pass so she could change the signals and retreat back inside. In the dark car, Michelle wasn't seen.

While other members of the delegation found the clacketa-clacketa of the wheels on the tracks almost hypnotic as they drifted off to sleep, the sound only reminded Michelle of the miserable, twenty-five-hour train trip she had made in 1988.

They had been traveling from Moscow to Volgograd, and she was sick. She had picked up an intestinal bug (referred to among the group as "Stalin's revenge"), perhaps in Leningrad the day before, and time and time again she had to dash to one of the suffocatingly small and smelly bathrooms at each end of the car. Their delegation just about filled the car, and everyone knew to clear the way when she ran by. The attendant in their car was aware she had a sick passenger, so she splashed disinfectant all over the bathrooms—which made them even more unbearable.

The worst part was that the bathrooms locked automatically when the train stopped, which it did often, and she would have to sit in her cabin playing mind games to convince herself that she would not be sick while the bathrooms were locked. She survived the train trip to Volgograd (despite declining Oleg's offer of a tumbler full of vodka as a sure remedy), but from then on, trains never held the romantic allure for her that they seemed to hold for others.

In fact, she hadn't fared too well with trains on the current trip either. Oleg had invited her to spend a night with him and his wife, Alyona, at their tiny apartment in Tscherbinka, a Moscow suburb. The only way Michelle could fit it into the delegation's itinerary was for her to come back by herself a day early from Vladimir on the electric train. Oleg would meet the train in Moscow and they would take his usual commuter routine of Metro, *electreechka,* and *avtobus* to the apartment.

Although the *electreechka* from Vladimir was jammed with Muscovites returning from the Revolution Day holiday, Michelle spotted Oleg waving to her as she jumped off the train in Moscow. They chatted on the hour-and-a-half commute, but when they got to Tscherbinka, a freight train blocked the way between the station and the bus stop. The night had come on quickly, and a light, wind-driven snow was falling. The freight train didn't present much of a problem for most of the commuters from the *electreechka;* they just stepped up between two of the cars and climbed over the coupling.

Michelle started walking along the train to go around when Oleg said, "That will take too long. We'll miss the bus. We'd better climb over."

Her mouth dropped open. She couldn't imagine climbing over a freight train under any circumstances in the United States, but—*when in Moscow, do as the Muscovites do.* She was no wimp! She handed Oleg her bag, slung her purse over her shoulder, hitched up her long skirt, and stepped into the gap between the cars. She got a good foothold where her leather boot wouldn't slip, pulled herself up onto the coupling, and climbed to the middle.

Then, "ca-ROOOMPF!" The train started to move. She was stunned. Later she told friends she had a very clear image of this blonde, forty-ish *Amerikanka* in a red knit hat clinging to the back of a freight car for dear life as it rattled along to Leningrad. Oleg was shouting for her to get off. Other commuters who were planning to follow her over the coupling were yelling in Russian, and she was analyzing the situation. It was clear she could not jump free of the tracks from her current perch—she was amazed at how wide the freight car was; she would have to jump down and then out.

So she jumped—down between the cars, and then she fell backward, away from the tracks just as the cars moved on. As Oleg pulled her up from the snow, her light blue coat bore the evidence of her close call: a wide swath of greasy dirt from the freight car was smeared across her chest, as well as on the backs of both sleeves where she had pulled her arms in to clear the train.

She and Oleg just held each other for a couple of minutes, and then she broke into nervous laughter. What a story she would have to tell this time! Oleg said all he could think of was trying to explain to Michelle's husband, Laurie, that he had let her get run over by a train in Tscherbinka. They were late when they got to the apartment, but they had quite an excuse for Alyona.

4

FULFILLING
THE PROMISE

STILL ANOTHER American-Byelorussian bond was deepening along with the autumn cold. After the Citihope team had returned to New York City, PJ Moore discovered he could not put the haunting face and pale blue eyes of Natasha Ptushko out of his mind. He began recording his thoughts to her on cassettes and sending them off, along with Christian rock music and other small gifts he hoped would lift her spirits and reinforce her will to live.

"Listen to the tapes and know that I am listening to them too. In a way, we are listening together," he wrote in one note. "Take the vitamins and eat the dried fruit. You have to get stronger so that we can visit longer when I come back to Minsk to see you soon. Perhaps I can come for the Orthodox Christmas on January 7."

Paul Moore and Michael Christensen hadn't forgotten Natasha or the challenge Dr. Olga had issued to them: "One thousand bottles of methotrexate," she had said, would be enough to convince her of a God of miracles. By the time they were back in New York, they were committed to the children of Chernobyl as the primary effort of Citihope's new international outreach, and to Dr. Olga's methotrexate challenge in particular.

Emergency medical relief was the immediate goal. For the long term, Citihope committed itself to personal advocacy for two or three children of very special medical needs—including the effort to arrange a liver transplant for little Antonina Meshkova at Children's Hospital of Pittsburgh, and perhaps a bone-marrow transplant at Cornell University Hospital. At the same time, they would be seeking hospitals and research centers specializing in childhood cancers and leukemia to train a group of Byelorussian doctors in Western protocols and techniques in treating radiation-related diseases.

Another effort would be to arrange rest-and-recreation vacations in the United States for sick and at-risk children of Chernobyl. The experiences the previous summer of Hole-in-the-Wall Gang Camp in Connecticut and the American Cancer Society/YMCA camp in Michigan, which they had first read about, would provide models from which to work and places to begin.

Paul started to work immediately on the search for affordable methotrexate, calling on the long list of past Citihope resources as well as any other leads he was offered. Telephone calls to pharmaceutical companies kept turning up the same responses: "We only deal with a specific list of agencies. You're not on that list."

Unaccustomed to taking no for an answer, Paul asked, "Who *is* on your list?" Among others was Interchurch Medical Assistance, Inc., in Baltimore, Maryland, which sounded like an organization that would be sympathetic to Citihope's appeal. He contacted IMA's director, a United Church of Christ minister named Paul Maxey, and learned that he was responsible for obtaining medicine for the missionary relief efforts of a number of nonprofit and church agencies.

Paul convinced Maxey that the children of Chernobyl project was a good fit for their purposes, and asked about the availability of methotrexate, specifically the methotrexate produced by Lederle Corp., which Dr. Olga had been trained to use. "We can probably get it for about ten dollars a vial," was the reply, approximately five percent of the two hundred dollar retail cost.

Now Paul took his children of Chernobyl appeal on the air with his weekly radio broadcast in New York City. With his deep, commanding voice, he began:

> *Pretend you are a mother whose child is sick. You live in Brooklyn. You can't get the medicine to save your child's life, but your friend can get the medicine to save her sick child! She lives in Manhattan and all*

the medicine is in Manhattan. No one in Brooklyn is allowed to buy any. Your friend's child is getting well, but your child is getting worse. There's no hope. Soon your child will die. . . .

Sounds like a bad dream, doesn't it? Well, it's happening right now, not in Manhattan or Brooklyn, but in the Soviet Union's Republic of Byelorussia, near where the terrible nuclear disaster of Chernobyl occurred in 1986. . . . Heartbroken parents are watching their little ones weaken and die. . . . Children who were barely toddling in the spring of 1986 would be entering school today if they had the strength to go to school. Instead, they lie in hospital beds literally wasting away while competent, caring nurses and doctors watch helplessly. No medicine. No hope.

The tragedy of tragedies is that if these children lived in America they could be cured. Here in America, eighty-five percent of childhood leukemia cases are cured using common but expensive oncological drugs like methotrexate. But in Byelorussia, they don't have methotrexate. The Soviet Union doesn't manufacture it and hospitals can't afford to buy it from the West. . . .

Dr. Olga needs our help. The children are counting on us coming through. You can participate in literally saving the life of a child of Chernobyl. [Then Paul played a tape of part of his interview with Dr. Olga.]

Methotrexate costs two hundred dollars a vial. Through Paul Maxey and Interchurch Medical Assistance, we are able to obtain this miracle drug for just ten dollars a vial. That's five cents on the dollar! It takes about five to six treatments over a year's time to effect a cure. So for fifty dollars, you can treat a child for a year.

Please join Sharon and me in what we're calling "Project Open Door"—miracle medicine for the children of Chernobyl. If you'd like to be part of a miracle, just call us right now. Every fifty-dollar gift will be used to help save a child. Thirty-seven thousand children are waiting to be cured—just through the Open Door. Perhaps one is waiting for you.

The Citihope phone bank lit up and stayed lit with people pledging fifty-dollar gifts to help. Soon, enough was raised to pay for Dr. Olga's one thousand vials!

IMA was able to come up with eighteen hundred vials of methotrexate for Citihope. "We also got themodium, thyroid-enhancing medications, and vitamins—in all, about half a million dollars' worth of medicine," Paul beamed.

By November 1, they had enough money raised and medicine in hand to send a shipment to Dr. Olga. "Why wait for Christmas?" Paul decided. "Let's send it for Thanksgiving."

Why not, indeed? What could be more American than taking a traditional Thanksgiving meal of turkey, dressing, and pumpkin pie to the children of Chernobyl, along with the medicine purchased through IMA and paid for with a ten-thousand-dollar grant from World Vision, as well as individual donations from Citihope radio listeners?

Citihope turned to the Byelorussian ambassador to the U.N. for help in pulling the effort off in just three weeks. Gennady Buravkin had issued one of the initial appeals that summer for help, and he was ready to lend his assistance to this typically American "do-it-now" response.

PAUL'S WIFE AND CO-HOST of the weekly Citihope radio show, Sharon Moore, volunteered to lead an all-woman team on the Thanksgiving trip to Minsk. Other team members included the producer of the Citihope radio shows; a TWA flight attendant and friend of Citihope; and a young woman who worked with inner city children at a church in Brooklyn.

The four women boarded an Aeroflot flight to Moscow from JFK International on November 17, with twenty-eight cartons of food and medicine worth nearly half a million dollars. Thanks to the ambassador's appeal and Sharon's persistence, the shipping charges were waived, and Dr. Olga's methotrexate was on its way to the Children's Hematological Center. Sharon's first stop in Minsk was to see Dr. Olga there.

"You said to my husband Paul that you needed methotrexate. He asked you, 'What would be a miracle? One hundred vials?' You said, 'One thousand vials!' Do you remember?" Sharon asked as they stood in the cold, dimly lit hallway outside Dr. Olga's office.

"Yes, it's true," the doctor replied as she studied the woman in front of her. In her late forties, Sharon was more than a foot shorter than her towering husband. Her auburn-tinged, dark hair was cut in a short bob, and it framed a round face that could be taken for Slavic. Her intense blue eyes tended to puddle up whenever something touched her—a fairly common occurrence on this trip.

"What would you say if I told you I had in the van right now eighteen hundred vials of methotrexate? Is it a miracle?" Sharon asked.

"Oh, yes, what a miracle! Eighteen hundred vials? I want to see it!"

As the van was unloaded, Dr. Olga started to cry. She hugged Sharon. "This is the greatest gift for all my children and for me also."

Sharon was in tears, too, as she climbed back into the van. "What a gift to be able to give a mother back her child."

Citihope was particularly pleased with the turkeys donated by the United Parcel Service. "Imagine terminally ill children and their families enjoying a feast in a country where food is scarce," Michael told the press back in the United States, "and all of them finding hope in knowing that medicine previously unavailable has been brought to them as a gift." The *San Francisco Examiner* ran a story November 23, headlined: "It Will Be Thanksgiving Day in Chernobyl—Food and Medicine on the Way," which was picked up by a wire service. It was Citihope's first national publicity.

On Thanksgiving Day in Minsk, volunteers were called upon to roast the turkeys according to Sharon's instructions, but the huge birds had to be cut in half to fit inside the smaller ovens in the Soviet Union. Two hundred and forty children (mostly orphans) were served complete meals, and enough remained to provide leftovers for a week—another American Thanksgiving tradition.

Because much of the meat and produce available in Byelorussia has been exposed to dangerously high levels of background radiation, the food from America had more significance for the children in Minsk than merely as elements in an American holiday feast. The meal represented their first in more than four years that was guaranteed to be free from radioactive contamination. That psychological boost, beyond the nutritional value of the meal, may have been enough to bolster those depressed immune systems a bit.

Of course, the feast was more than just a meal. The children put on a variety show and sang folk songs for the American visitors. Sharon told them the story of the first American Thanksgiving in 1621, with special emphasis on the idea of thanking God for supplying needs.

During the celebration, Vladimir Lipsky of the Children's Fund told the children, "This is your first Thanksgiving in the Soviet Union. When you grow up and take power in this country, you must also have a Thanksgiving so that your children will know how to give thanks to God for food and medicine."

* * *

AFTER THE CELEBRATION, the four women boarded a tiny, propeller-driven plane of questionable safety for the contaminated regions within the thirty-kilometer danger zone around Chernobyl. They wanted to bring some of the medicine and medical supplies to hospitals and orphanages in the Gomelskaya region, where it was needed the most.

They were told they were the first Americans to be taken past the second restricted ring around Chernobyl in the "dead" thirty-kilometer zone. There they saw firsthand the desolate, evacuated villages. Stopping at one of the small cities on the perimeter of the zone, Sharon met a scientist who handed her a bootleg videotape of the secret liquidation efforts conducted in the first few hours after the explosion. Citihope later used the tape on television to help tell the story of the children of Chernobyl.

In the city of Gomel, the largest city receiving a deadly dose of radioactive fallout, Sharon and the other women spent the day with children and mothers in a hospital and an orphanage. They were overwhelmed and emotionally drained by what they saw. Like those who had gone before—the earlier Citihope team and the UCC delegation—they felt powerless.

At one stop, Sharon sat on the floor with a girl of eleven or twelve and asked what it was like to be a child in Gomel. "I admit that I am afraid of the radiation, especially when it's raining. When I stayed with my friends at a sports camp, it was raining and we all wanted to cover ourselves and hide. We know the rain is very dangerous."

Sharon turned to her interpreter. "This is no way for a child to live."

"Of course, you are right," the interpreter replied. "There is a feeling here that Byelorussia, in some future period of time, will be absolutely dead."

Once she was back in Minsk, Sharon inquired at the Oncological Hospital about some of the children PJ had videotaped on the earlier trip. The chief doctor told her: "There is a saying that people usually die in the late fall and the early spring. In October we witnessed one child dying every day for a week."

As if driven by the doctor's words, Sharon made her way through the hospital wards, anointing the children and praying for healing. As she put the oil on the brows of the children, mothers would often motion for her to anoint them as well. Some would pray with her. Others would press copies of their child's medical records into her hands and plead with their eyes. No words were needed.

❀ ❀ ❀

SHARON HAD one final mission before she could leave. She was carrying photographs and gifts for Natasha Ptushko from her son PJ. As she was led into Natasha's room at the Oncological Hospital, Sharon found the young girl with the luminescent eyes so weak she could not sit up.

"Hello, Natasha. *Menya zavoot* (My name is) Sharon Moore," she said in carefully rehearsed Russian. She took Natasha's thin hand in hers and held it close to her face. She wanted to get to know this young woman that her son had grown to love.

"I know you are the mother of Paul Moore Jr.," Natasha whispered weakly.

"I woke up very early this morning and my first thoughts were of you," Sharon said. "My first prayers were for you, that God would give me the words to communicate his love, his power. I'm not a healer; I have no power to make you well. The only power I have is God's, for my life and for yours."

"For those who pray for my healing, you can't imagine how thankful I am," Natasha responded in a voice as thin and frail as her body. "When you entered my room just now, I felt you were the kindest person I'd ever met. You're like a mother to me. How lucky I am that you are here. I haven't slept at all since they told me you were coming. I have just been waiting for you."

On the table next to her bed was a small American flag that PJ had given her. Sharon opened her bag and took out two small picture frames—one held a photo of PJ in the uniform of his military school, and the other, a picture of Jesus Christ—and set them on the table.

"When you go to bed at night, before you close your eyes, you can look at the faces of two men who love you—one with earthly love and one who loves you more than any man on earth is able. And when you wake up in the morning, you will know that Jesus has held you in his arms all night long, to give you strength for the day."

When words failed, Sharon took out a small vial of oil and anointed Natasha in the name of the Father, Son, and Holy Spirit. She whispered the ancient words in Latin: *"Kyrie eleison, Christe eleison;* Lord have mercy, Christ have mercy."

"Will your son come in January? He said he might," Natasha asked with obvious hope.

"Yes, he will come at Christmas."

"Then I promise I will not die before he comes."

❋ ❋ ❋

ONCE SHE WAS BACK in New York, Sharon told a responsive audience at the Lamb's Theater in Times Square how it felt to visit abandoned villages inside the thirty-kilometer zone. "The deafening silence of totally evacuated villages was eerie. The wind blowing through the ghost towns reminded me of an Alfred Hitchcock film. Visiting the empty kindergarten in one village, we saw children's artwork, books, and toys exactly the way they left them when the mass evacuation took place.

"Holding a child's clay sculpture of a basket of berries made me wonder where these children are now," she said. "We were told that at least forty percent of the residents of this village are now dead. The rest are being monitored or treated for radiation sickness and other chronic diseases."

Among the slides that Sharon showed that night was one of Natasha in which she looked very much like Michael's image of "the white rose." When Pyotr Kravchanka, the Byelorussian minister of foreign affairs, got up to speak, he too was moved by that image.

"I am impressed by your pictures of the girl named Natasha. What is her fate? Will she grow from a young girl into a woman? Will she become a wife and mother? Will she give birth in the future? Or will she wither?

"We don't know, but we believe your charity and assistance and the warmth of your hearts will save Natasha. Hope is the name of this evening. May this evening of hope truly be a beacon of assistance for hundreds of Byelorussian children."

The foreign minister thanked the "courageous American women who ventured beyond the ocean to the unknown world on a mission of understanding. One year ago, the Berlin Wall fell, but something more than the wall fell. That which prevented mutual understanding and compassion fell as well. . . . We start to understand when we see each other better."

Earlier that month, Kravchanka had presented a progress report on the children of Chernobyl to a committee of the UN General Assembly. He began with special thanks to the UN disaster relief coordinator, who had called for an international response to the disaster. Then he continued: "Without exaggeration, the situation in Byelorussia is extremely grave. . . . Only recently we have received the latest information from the Academy of Sciences of Byelorussia, according to which a new series of radiological measurements has revealed much larger scales of radioactive contamination in the republic than we had believed them to be quite a short time ago."

He went on to say that five out of the six regions of Byelorussia had been affected by the contamination, including a large portion of the Minsk region, which had earlier been considered to be relatively free of it. "Only one percent of the republic's territory remains clean," he continued.

"Moreover, the research reveals a number of new, previously unknown post-Chernobyl factors that negatively affect the health of people and the environment. This unfortunately convinces us that mankind—as in the well-known surrealistic painting by Salvador Dali—is just pulling up a bit of the brim of the sea which hides underneath yet unknown consequences and dimensions of the Chernobyl crisis."

Kravchanka drew a poignant portrait in closing. Noting that many cities throughout the world have digital time and temperature signs as part of the urban landscape, he added softly, "In Minsk, such a display now shows a third measurement—the level of radioactivity. Evidently, it will stay with us for many years to come."

Starting now, he implored, we "wish the clocks in every home to begin to count a new time, a time of broad international humanitarian solidarity and mutual assistance for the sake of the children of the future.

"Mankind is strong only when the other nation's grief is felt as one's own, when invisible but strong ties of common humane and cultural ideals unite us. Following the great Flemish author, we say: the ashes of Chernobyl are burning into our heart; the ashes of Hiroshima are burning into our heart; the ashes from all nuclear explosions that lie heavily on the earth are burning into our heart."

* * *

BEFORE THE YEAR ended, an unfinished project demanded Michael Christensen's attention—the task of convincing Children's Hospital of Pittsburgh to provide a liver transplant for Antonina Meshkova, the toddler who continued to defy the predictions of her doctors and fight for life. Michael's initial telephone calls to the department of surgery at the University of Pittsburgh School of Medicine established some very disturbing facts. "The average cost of a liver transplant in the United States is $225,000. It could go as high as $800,000," he wrote in his notes. "The hospital is unwilling to consider the operation without a substantial deposit and commitments to pay the fee."

Michael followed up with a personal plea in a letter on November 8: "My simple request is that you would help find a way for Antonina to receive the liver transplant she needs to live.

"You indicated that the only impediment to treatment would be 'administrative,' that is, financial. You suggested that the national health program in the USSR simply contact the administration of Children's Hospital in Pittsburgh and make the necessary arrangements for a deposit.

"Surely you realize that Soviet rubles cannot pay for such an operation in the United States, and neither the Soviet national health insurance program nor the Byelorussian Children's Fund has enough hard currency to pay the entire bill."

Michael contacted a leading surgeon at the hospital a few weeks later and obtained his commitment to waive his fee of about $45,000. That at least was encouraging. But the hospital bureaucracy would not allow Citihope to open an account for Antonina at the hospital until half of the necessary $165,000 deposit was in hand. The administrator added, "There are twelve Russian children and their families requesting transplants here. If we accept one, we would have to accept all of them."

"I'm not asking you to accept all twelve," Michael pleaded over the phone as he pushed the door to his office closed. He worked out of his home, and the sounds of his daughter Rachel playing in the next room were making it hard to concentrate. "I'm only asking on behalf of one. I'm praying you will adopt the attitude of Mother Teresa, who says, 'I am only one, but I am one. I cannot do everything, but I can do something.' "

Several more telephone calls followed, but the result remained the same: Antonina would not be getting a liver transplant in Pittsburgh any time soon.

Michael sat with the receiver in his hand until the dial tone buzzed. Slumping into his chair, he rubbed his hands over his face. He had failed a spunky little girl, just a few months older than his own Rachel. The weight of this realization was suffocating. He and Paul had begun this project intending to make a positive difference in the lives of these children. "And we have—." His voice came out in a whisper. "I know we have, for hundreds of them."

He couldn't help Antonina, but what about the others? Slowly and deliberately, he raised his head, pulled himself out of the chair, and regained his composure. He opened the door of his office, called to Rachel, put her on his lap, and said, "Why don't you help Daddy make the next call."

PART TWO

In the still, brittle cold of the late afternoon, a steady snowfall adds a fresh coat of brilliance to the sooty crust beneath. Snowflakes cling to the branches of the bare birches and the ever-dressed firs. It's novee got, New Year's Eve, *in the medieval city of Minsk.*

At the city's edge, where the rows of sterile, cement-block apartment buildings are just a bump on the horizon and occasional saplings are the forerunners of the forest's attempt to reclaim the clearing, the shouts and giggles of children compete with a scratchy, over-amplified recording of Prokofiev's "Peter and the Wolf."

Fathers, with cigarettes clenched between nearly blue lips, stomp their boots in the snow and dig their hands deeper into their pockets. Mothers help their youngest ones lace up skates and hold tight to steady them as they stagger onto the ice pond at the edge of the woods.

Teenagers speed across the ice with hockey sticks poised, and the littlest skaters, barely mobile under layers of sweaters, snowpants, and mittens, are bumped and sent sprawling.

The babushka, *who plays the records and sells paper-wrapped* piroshkee *from her shed at the edge of the frozen pond, shakes her fist at the teenagers, but they don't even glance her way.*

After all, it's novee got *and Grandfather Frost comes tonight!*

5

PROJECT MAGI

PROJECT MAGI was the code word for Citihope's twelve-day mission of mercy at Christmastime to the children of Chernobyl. The name was intended to recall the story in the gospel of Matthew of the magi, the wise men from the East, who followed a new star in the sky to find the newborn Christ child in Bethlehem and honor him with gifts of gold, frankincense, and myrrh. In this modern version of the story, however, the "magi" were coming from the West, following a star of hope and seeking the Christ child among the children of Chernobyl. The gifts they brought were toys, antibiotics, methotrexate, and uncontaminated food, to be delivered on Orthodox Christmas Day, January 7.

This would be the first officially sanctioned observance of the sacred holiday of *Rozhestvo* since Josef Stalin erased Christmas from the Soviet calendar in the 1930s and replaced it with a secular winter holiday celebration focused on New Year's Day. Russian Orthodoxy still adheres to the ancient Julian calendar instead of the later Gregorian one, and Christmas in Byelorussia comes January 7 instead of December 25 as it does in the West.

During the years of official state atheism, *Rozhestvo* was only resurrected once—at the outbreak of the Great Patriotic War (World War II). The state needed the church to rally the people for the defense of the *rodina,* the motherland, against the onslaught of the hated fascists. Stalin called a halt to the persecutions of the church and ordered his lieutenants into the *gulags* to find surviving priests who could issue a call to arms. The church enjoyed a fitful flourish and then once again was pounded into submission by the atheistic fury of Nikita Khrushchev.

But now, as the seven Americans from Citihope arrived in the Soviet Union in December to launch Project Magi, it once again became legal to observe Christmas—in fact a state holiday had been declared—and to tell the story of the nativity of Christ to a generation of children who had never heard of the Babe of Bethlehem.

This is what had motivated the Americans to give up their own holiday and embrace a new one in the USSR. The delegation included Michael Christensen and his wife, Rebecca Laird-Christensen (they had left their nine-month-old daughter with grandparents and were suffering through their first separation from her); Greg Schneider and Beth Lueders, who were covering the story for *Worldwide Challenge* magazine and who would meet the other five in Minsk after a stop in Poland; and the Moore family—Paul, Sharon, and PJ—who had become celebrities of sorts in Byelorussia.

The current political situation was uncertain at best. The United States had just given Iraq an ultimatum to get out of Kuwait by January 15. The country was poised at the edge of war in the Persian Gulf when the group left New York for the information vacuum of the Soviet Union. They took an Aeroflot/Pan Am flight to Moscow four days after Christmas, with eighty-six boxes of relief supplies weighing nearly six thousand pounds.

Michael and Rebecca connected with the Moore family at JFK airport in New York.

"You look like you're coming back instead of leaving," Sharon said as she hugged each one. "You must be exhausted."

"We are. Rebecca just finished her book manuscript and I gave three radio and television interviews on the trip. Then we kissed Rachel goodbye and climbed on the 'red eye' for five sleepless hours so we could get here at dawn," Michael said.

"Well, you can sleep on the plane," Paul said as he clapped his long arm around Michael's shoulder. "Let's see if we can get all these boxes of drugs and medical supplies on the plane with a minimum of hassle—for a change."

United Parcel Service had supplied a semi-truck and trailer to haul the supplies to the airport, but the Moores were learning that moving cargo was never simple. Aeroflot had agreed to load the eighty-six cartons as "extra cargo" without overage charges only after Sharon had cajoled, pleaded, and shamed the station chief of the Soviet national airline into helping the USSR's own children of Chernobyl.

"We may have to sic Sharon on them again," Paul said with obvious pride. "You know she can be a real tigress when she has to. In dealing with these mid-level Soviet bureaucrats, she's developed an effective ritual. First, she asks politely; then she pleads with them to do the right thing. If that does the trick, she gets it in writing and thanks them for their decisiveness and spirit of cooperation.

"But if she comes up empty, then sweet, wide-eyed little Sharon is not above threatening to tell the world through the media how cold-hearted the Soviet national airline can be in providing for its own children.

"That usually works. But if it doesn't, she finally will plant herself on top of one of the cartons and refuse to leave until the bureaucrat comes to his senses. So far, one of her strategies has always gotten the cargo loaded."

On this December day, Sharon was pleased to find a familiar face in charge of the Aeroflot cargo department. She had developed a cordial relationship with this agent over the past few months and, perhaps moved by the spirit of the season—or the lack of between-the-holidays travelers—he agreed to take the cargo free of charge. No pleading or intimidation required.

"That will save us a tidy eight thousand dollars," Paul calculated. "That means we can add about four hundred vials of methotrexate to the next shipment."

While Sharon was finishing her negotiations, Michael was doing a little deal-making of his own. He managed to get the five of them upgraded from coach to the obvious comforts of the first-class cabin.

"How else were we supposed to catch up on our sleep?" he crowed as he handed new boarding passes to the rest of the group.

Spirits were soaring when the Laird-Christensens and the Moores settled into those wide, pseudo-leather seats for the eight-hour flight to Moscow. The realization that they would share in the first officially sanctioned *Rozhestvo* celebration in the Soviet Union in generations had each of them giddy with excitement.

Arrival at Sheremetyevo Airport in Moscow, however, brought them down to earth in more ways than one.

"I don't see anyone that looks like the English-speaking representative of the Byelorussian Children's Fund who's supposed to meet us," Sharon said as she scanned the nearly empty passenger terminal. "The fax said they would arrange for the cargo to be unloaded directly from the plane to their truck."

"That may have been the plan worked out by telephone and fax before we left New York," Michael said, "but, as usual, it looks like nothing is as simple as it seems. I think the word *snafu* was coined for any attempt to work out details at a distance with anyone in the Soviet Union. Has anything ever gone according to plan on any of the past trips? I don't think so."

There was no BCF representative, and definitely no truck of any kind. The five Americans watched helplessly as all eighty-six boxes, weighing about seventy pounds each, glided into the baggage claim area on the conveyor belt along with the personal luggage of the Pan Am/Aeroflot passengers. All five were exhausted, despite the first-class comforts of the trip.

"Well, you can stand here all day and hope that someone turns up," PJ said as he picked up one of the cartons, "but I'm going to start moving these boxes. They may just decide it's all unclaimed baggage and lock it up somewhere."

So, box by box, the three men and two women unloaded the three tons of cargo that kept parading around and around on the conveyor belt. For the next few hours they piled it into a cardboard mountain in front of the customs declaration desk—to the obvious bewilderment of the customs officials.

"I'm beginning to think that no one is coming," Sharon said as Paul put the last box in place. "We've got to get all this through customs, and I don't intend to pay for the privilege of bringing medicine to dying children in this country. Let's see how far the Russian phrases I've learned will take us."

Sharon walked up to the bored and somewhat amused customs officials and tried to convey to them that those eighty-six cartons contained medicine and medical supplies for children of Chernobyl and they should be eager to allow them into the country. However, one bureaucrat after another appeared on the scene to shake heads, shrug shoulders, or wave hands to a chorus of "*nyet.*" It didn't appear that Sharon's tigress routine worked nearly as well on the Soviet end of the cargo chain.

This was Rebecca's first trip to the Soviet Union, and this strange and exotic land was not making a good first impression.

She sat down on a carton of antibiotics to make her first entry in her journal:

> *Moscow's drab and unfriendly airport was in a rural area surrounded by tall trees. A snow-scattered tundra with a smattering of houses sitting on uneven and roughly fenced parcels of land lay beyond the confines of the airport.*
>
> *Military service must have been required as pale, pimply faced boys, teetering on the edge of manhood, staffed the entry points. A wink and wide smile were bestowed on this American lady.*
>
> *Was pot smoking legal here? The sweet, unmistakable odor of marijuana smoke wafted through the stale air from time to time. Everyone seemed to be smoking, and there was a run on duty-free beer, which was consumed as passengers waited for luggage.*
>
> *Few people smiled. They rudely guarded the baggage carts for their own use and pushed or bumped as they passed by. No one spoke or offered to help. Hard times seemed to have exacted a toll on the reservoir of kindness here.*
>
> *Two elderly women dressed in drab work smocks pushed garbage cans through the airport. One shuffled along in worn bedroom slippers. I would bet they went home at night to drab, colorless, cramped rooms after fighting lines for rationed goods. Life seemed hard. The underlying depression was palpable.*

Rebecca, who was in her early thirties and a professional writer, editor, and licensed Nazarene minister, had now met the public side of the Soviet people—the cold, brusque side that offered shoves or shrugs instead of smiles. It would take longer for her to know the private side that could envelop a brand-new American friend with warmth, hospitality, and love.

Three hours later their local host arrived wringing his hands.

"I'm so sorry about the delay but the van broke down and I could not hire another one. I had to fix it myself," he said as he ogled the mound of cartons.

"But now I see that we have another problem. No one told me that you would be bringing all this with you. My small van will never hold all these boxes. I'm afraid you must wait again while I try to hire a truck."

It was two hours more before a truck—actually a bus—was commandeered to handle the cargo. In the meantime, the interpreter/guide—short, squat, and reserved—proved to be persistent and finally successful in helping the group clear customs with all the cargo—and with no additional fees.

Now there were seven (including the bus driver) to haul the eighty-six cartons onto the bus, across an icy parking lot at dusk in the bitter cold of a Russian winter day. It wasn't long before Rebecca had strained her back, so the hauling fell mainly to Michael, Paul, and PJ. Before the day was over, the three would have manhandled six thousand pounds of cargo on four different occasions.

The Citihope five and the cargo still weren't on their way to Minsk, though. They were just driving across Moscow to the domestic airport, where they would catch a flight to the capital of Byelorussia. At the domestic airport they could look forward to another five hours of negotiations.

As Sharon and the interpreter argued with the bureaucrats, the others stayed with the cartons on the bus in subfreezing temperatures. Each one had his or her own way of coping: Paul kept getting on and off the bus, unable to sit and wait patiently for a solution; Michael, tired and depressed, sat quietly near the bus's heater, bundled up and zoning out; PJ put on his Walkman headphones and escaped into his music; Rebecca pulled out a book and read until she dropped off to sleep.

Finally, the answer came. No problem getting the five bone-weary travelers onto the flight, but the plane was much too small to take the cargo. The solution was quite simple: the Americans would take the plane to Minsk and the boxes would get there when a larger plane could be chartered.

Sharon was livid. "No way am I letting this cargo out of my sight! I made a sacred promise to the donors that I would personally supervise the delivery of this medicine to the children of Chernobyl, that I would not let it be warehoused or diverted into the black market. I will stay with these boxes no matter how long it takes to get them to Minsk."

She didn't stamp her foot; she was just too tired.

Nothing could convince her or Paul that the cargo would be safe without their physical presence, and the five finally agreed that one would stay while the rest took the flight to Minsk that night. Rising to the role of martyr or, at least, servant leader, Paul made the decision to stay with the cargo. "I'll stay and raise Cain here until they send a bigger plane. You do what you can in Minsk to get what we need."

Sharon finally agreed and kissed him goodbye. The other four dug into their packs and produced trail mix, granola bars, and chocolate—"little condolences for our inconveniences," Paul called them. Michael gave him his emergency space blanket—a tube of fortified aluminum foil that retained heat—and wished him well.

The remaining four squeezed themselves into the cramped confines of the tiny and obviously ill-maintained propeller craft. They took the first row of seats, which faced another row of seats with tables in between. With no personal space or room for their feet, the Americans were uncomfortable and anxious. The man across from Rebecca had placed his hand luggage on the table, with some poles sticking out at right angles. Another bag under the table forced her feet into a narrow crevice. So much baggage was packed around Sharon's legs that her feet went to sleep. The water dripping from the ceiling created some added concern about the airworthiness of the plane. Fortunately, the flight was only an hour long, and the most-welcome touchdown suggested that planes just like this one probably went up and came down across the Soviet Union every day—usually without mishap.

It was after midnight when the four tired and grumbling Americans were greeted with warmth and enthusiasm by friends from the two earlier trips—Vladimir Lipsky and Alexander Trukhan, both of whom presented red carnations to the women. Bubbly and blonde Nadia Vyalkova, who had served as an interpreter on the Thanksgiving trip, hugged Sharon like a daughter greeting her mother after a long absence, and an Intourist aide served them hot tea and coffee as they waited for their baggage.

The sagging spirits of the Americans rebounded as they responded to the warmth of the welcome that deep winter night in Minsk.

❊ ❊ ❊

BACK AT THE DOMESTIC airport in Moscow, nervous officials assigned two young militiamen to help Paul guard the cartons, which were stacked floor to ceiling in the passenger terminal (and no doubt to keep an eye on Paul as well). Paul shared his snacks with the guards and tried to communicate with them through pantomime and pointing, but they kept leaning their heads on their hands, pressed palm to palm, to indicate that he should go to sleep.

Finally Paul could stay awake no longer. He wrapped himself in Michael's foil blanket, squeezed his six-foot, four-inch body under the armrests of a standard airport bench, and settled in for a long, cold, and uncertain night. But sleep wouldn't come because he couldn't find a comfortable position, and every time he moved, his foil cocoon would rattle and a blast of cold air would chill him to the bone.

Sometime during the night, through the fog of his twilight state, he felt two very soft hands lift his head ever so gently, slide a folded coat under it, and then spread another one over part of his long body. He opened his eyes to see a smiling, middle-aged woman in the smock of an airport cleaning attendant. She patted him on the head and then shuffled away.

"I felt great comfort and fell fast asleep for the rest of the night," he later told Sharon. "I thought I'd been visited by an angel."

In the morning, remarkably refreshed, Paul found a small airport buffet open and bought some sausage and good Russian bread for a most welcome breakfast. Then he hired a couple of men he found nearby to help haul the eighty-six cartons one more time—this time to the large Aeroflot passenger jet that had arrived to carry him and the cargo to Minsk.

A regular flight to Minsk was scheduled for that morning, but someone had ordered a much larger plane that could accommodate all the cartons. The passengers filled the back of the plane; Paul had forty or fifty seats in the front to himself. He even had his own flight attendant.

In Minsk, Aeroflot and Byelorussian Children's Fund officials met him with copious apologies—and a truck. He didn't have to lift a single box another time on the trip.

After a glorious night's sleep in real beds, the American contingent in Minsk was delivered to the offices of the Children's Fund on Communisteechiskaya Street on the morning of December 31. There, at a conference table laden with Pepsi, mineral water, and sweets, they listened to an annual report from Vladimir Lipsky.

Smiling and as animated as ever, he began to tick off BCF's accomplishments. "Last year we sent 1,235 children to other, noncontaminated republics in the Union as well as to fifteen different countries in Europe and Cuba. We sent them so they could rest and enjoy themselves, just be children, without the risk of contamination.

"Then, in the summer, representatives from twenty-two countries, including the United States, met here in Minsk to talk about taking even more children to camps next summer. We are very much encouraged by the interest because, once a connection is made and these people meet our children, they are ready to help in other ways, too."

Michael made a note of the American groups he mentioned—the YMCA, American Cancer Society, and the Christian Children's Fund. Michael remembered reading about the YMCA efforts before, in that first *New York Times* article that had sparked Paul's initial interest.

Lipsky continued: "We've also created some new family services and orphanages as well as medical relief for the children of Chernobyl. And we've started two new magazines—*The Children* and *We and the Rainbow*—which I edit and the Fund publishes. We want to turn the attention of the people and the government leaders to the needs of children."

Paul wanted to hear more about medical services, since that was the area he expected Citihope would be able to support most directly.

"The Ministry of Health has joined us in sponsoring the Anti-Chernobyl Diagnostic Center. There we are monitoring precisely the health of eighteen thousand selected children of Chernobyl who live at risk daily in the contaminated regions of our republic.

"We don't regret our gray hair and wrinkles," Lipsky said in his familiar, effusive way, "for our hard work is all for the sake of the children."

By the time he was finished, his white hair fell over his forehead and his elfin face was flushed. Enthusiasm came easily to Lipsky, and the Americans were in the mood to respond now that they knew that the cargo was safe and dry in the BCF warehouse. Besides, it was New Year's Eve and time to celebrate.

❖ ❖ ❖

NOVEE GOT, the New Year's celebration in the Soviet Union, had woven some of the elements of the traditional Christmas observance, dressed in new secular garb, into a nonreligious winter holiday theme. The key figure was *Dyed Moroz* or Grandfather

Frost, who bore a striking resemblance to Santa Claus. He had a long white beard and wore a red velvet robe and hat; he arrived at the stroke of midnight to present gifts to good children who never complained, obeyed their parents, and never, ever fought with their brothers and sisters.

If the parents of Soviet children happened to be well-connected, it could be arranged for Grandfather Frost, usually an employee of the state, to make personal visits to individual apartments accompanied by *Snegochka*, the snow maiden—often a local beauty queen—who was also available for Grandfather Frost to lean on after he had been rewarded in each apartment by a glass of holiday vodka and an appropriate toast.

The centerpiece of a New Year's party was always the New Year's tree, a fir or pine decorated with colored glass balls, electric lights, and tinsel, and topped with the *krasnaya zvezda*, the red star of the Soviet state.

New Year's was a two-day national holiday when Soviet families gathered for good food (with menus adjusted to reflect what could be found in the stores or bartered for in the streets) and good cheer. But for those children of Chernobyl with no homes, institutions did their best to fill the gap.

The staff of the orphanage and boarding school in Minsk had told the children that Grandfather Frost was coming and he was bringing Americans with him. This was the same school that Sharon Moore's delegation had visited at Thanksgiving, where they shared the turkey dinners. More than forty children, ages nine to fourteen, could hardly contain their excitement. Their New Year's tree was decorated for the season, food was prepared, the table set—and finally the guests arrived.

After the formal greetings, which included a song-and-dance performance by the children and a response by Sharon, Paul folded his long frame onto the floor and brought out a mantlepiece crèche with ceramic figures. With the children gathered around, he acted out the story of the most glorious birth in the stable behind the inn as he put the pieces in their familiar places in the crèche. Rebecca paraphrased the narrative from the gospel of Matthew.

> *A long time ago, in a faraway land, three magi watched the stars. One night a star that was brighter than any other appeared in the East.*

The magi remembered that it had been foretold that a new king would be born. Maybe this was his star.

The magi followed the star to the town of Bethlehem. There, in a stable, they found a baby named Jesus lying in the straw. The star still shown brightly overhead. This was the new king—a different kind of king. The stable was filled with love, peace, and healing, which are the gifts he came to bring.

The magi worshipped this tiny king by giving him the very best gifts they had to give. One gave him pieces of gold; one gave him frankincense to burn; the last gave him sweet-smelling myrrh. All three gave him their hearts.

Each year at Christmas, we celebrate the birth of Jesus by giving gifts. We may not have gold, frankincense, and myrrh to offer, but we can all give our hearts to Jesus and share his gifts of love, peace, and healing with each other. Like the three magi, we can follow the star and find Jesus.

Merry Christmas!

When the crèche was completed and the story was done, Paul asked, "How many of you have heard the story of the baby Jesus before?"

Not one hand was raised.

"Then you must get to know him. Would you like to touch baby Jesus in the manger?" The children crowded around, each in turn extending a finger to stroke the porcelain figure.

Then it was time for the gifts. Wrapping was torn away to reveal singing clowns, toy cellular phones, cars, and Lego kits. Toys with musical components were obvious favorites. Amid the din of clanging fire engines and American rock music, several children recited poems and sang songs about the fir tree.

During an elaborate tea in an upstairs lounge, Rebecca tried to cut through the adult speechmaking to reach the children and their thoughts at this holiday season.

"Tell me your wish for the New Year," she asked each child. They were all shy, but with a little encouragement, she elicited some responses. One little girl shared her wish for world peace; another asked for progress in US-USSR friendship. Three giggling girls sitting on the sofa conferred and decided they wanted better living conditions in their country—and world peace, too.

One pale, wistful, fourteen-year-old boy sitting alone at the table began, faltered, and then found his courage as color rushed to his cheeks: "I wish for parents—and that all the boarding schools were closed because there was no longer any need for them."

The Americans were visibly touched by the depth of his longing. Sharon hugged him and said, "I may not be able to take you home to New York, but I will give you a home in my heart."

This episode remained with Rebecca and she commented on it that night as she wrote in her journal, "*It struck me that our well-meant promises of a heart-connected family were sincere, yet they would make very little difference to this young man's daily life. His chances of adoption were very slight. I was moved that this boy, already embarked on manhood, still wanted a mom and dad just at the age when most teenage boys wished they could be free of their parents.*"

❋ ❋ ❋

IT WAS NEARLY 10:00 P.M. on New Year's Eve when the Americans returned to the elegant complex where they were staying, in the woods outside Minsk.

Armed militiamen stood at the iron gates at the end of a long road cut through the forest of bare birches and snow-heavy firs. Their boots crunched on the sooty crust of snow as they opened the gates for the caravan of cars.

A prerevolutionary estate, perhaps once belonging to a Romanov prince or a Byelorussian *boyar*, the compound retained its pre-communist glory as it continued to serve as a showpiece for the *apparatchik*. Victorian lamplights illuminated the network of paths connecting the "cottages" to the mansion where the night's festivities would take place.

A personal staff hovered nearby as the Americans were escorted into the splendor of the dining room, with its many-tiered, gold-leaf chandelier strung with crystal beads, Persian carpets over parqueted floors, and panels of carved mahogany lining the walls.

After a steady diet of spartan hotel rooms and especially dreary hospital wards, the wealth of the furnishings left the Americans gaping. Paul Jr. leaned over to Michael and whispered, "Just remember, in the USSR, all comrades are equal, but some are more equal than others."

The hospitality was as splendid as the surroundings. Their hosts included Vladimir Lipsky, who had brought his daughter; tall and blond Alexander Trukhan; the very refined Pyotr Kravchanka (the Byelorussian minister of foreign affairs, who had arranged for the plane that had fetched Paul and the cargo); his wife; and Irina and Nadia, the interpreters. The gaiety of the holiday had infected everyone with an abundance of good will and good cheer. The agonies of the group's arrival already seemed like a distant dream.

However, these American church people found themselves on unfamiliar ground. For them, a little wine with dinner was the extent of their experience with alcohol. They were unprepared for the steady flow of vodka and champagne, and worried that they would offend their hosts by refusing.

While they were wondering, Lipsky lifted a champagne glass and stood to offer the first of an evening full of long and flattering toasts.

"I raise my glass to honor the friendship between the Byelorussian Children's Fund and Citihope. Those of us gathered around the table this New Year's Eve are joined together in an unbreakable bond of commitment to the children of Chernobyl."

Paul stood and returned the toast with equal effusiveness a few minutes later.

Just before dinner, Lipsky presented "Year of the Sheep" plaques to everyone, and then he told a story from his childhood:

> When I was a little boy, about one or two years old, I was living in the Gomelskaya region when the Nazis came. They burned our village. My earliest memory is living in the forest without a home. My mother would tell me, "Don't laugh, don't cry, don't talk or they might hear us and kill us." Three years later, we returned to our villages. I didn't cry because my house was burned. I cried because all my toys and dolls were gone. . . . So now, whenever I travel, in every city I'm in, I buy a toy. When I was in New York City, I bought my favorite toy—Mickey Mouse.

He finished his speech with another toast: "My motherland still flows with blood. Our hope is in the children."

Now it was Sharon's turn to offer a toast. Her speeches often sounded like monologues from a Victorian novel, but they touched her Byelorussian hosts in a way that normal American conversation would not.

Paul picked up on Lipsky's disclosure that the new year would be the Year of the Sheep. As only a minister would dare, he offered a toast in the form of a prayer to the Good Shepherd, the Christ, and Sharon invited the Americans to recite Psalm 23 from memory. The words of the psalm seemed particularly meaningful to Rebecca that night. Later she wrote in her journal:

> "*Yea though I walk through the valley of the shadow of death. . . .*"
> *Here we were, in a nation in despair, visiting children with lethal diseases.*
> "*Thou preparest a table before me in the presence of mine enemies. . . .*" *We sat together, Soviets and Americans, enemies by birth and by virtue of a Cold War we had not undertaken. Both sides acknowledged feelings of enmity and fear for the other, and now we sat at table feasting together on fish* en glace, *caviar, mushrooms, and torte made by Mrs. Kravchanka as well as a dozen other dishes prepared by the staff.*
> "*My cup runneth over. . . .*" *Our hosts filled our glasses with champagne, then wine, then vodka—despite our protestations of* "Chut, chut!"
> "*Surely goodness and mercy shall follow me all the days of my life.*"

Irina, the young computer linguist, interpreted the psalm with a smoothness and proficiency that indicated the words were familiar to her. "I already know the scripture," she explained. "I learned it when I served as an interpreter for the Slavic Gospel Mission when they visited Minsk not long ago. They were here as part of their journey to bring a million Russian Bibles into the Soviet Union."

The evening's most memorable toast came from Paul Jr., who raised his glass in honor of his new life in the coming year.

"I must tell you that I was a very different person when I arrived here in Byelorussia last fall. I had been brought up to believe that the Soviet Union was the Evil Empire. I carried a lot of excess baggage with me on that trip—mostly hate for America's former enemies. Communists were the bad guys and I couldn't imagine any other way of seeing them. I really didn't even want to come and I didn't try to hide the way I felt. I think I was looking for an excuse not to help the children.

"Then I met Natasha Ptushko and my life was changed. I came face to face with the children of Chernobyl. I learned to love my enemy, but even more important, I learned you weren't really my enemy and you probably never were.

"I'm not very proud of the person that I was. But I'm a new person now and you have enlightened me. I love you all and I thank you for teaching me what it really means to love—and to live—again."

After an unusual few seconds of silence when PJ sat down, an obviously moved Kravchanka responded: "You are our hope for the new year and for the future."

Despite the upbeat mood of the evening, the interpreter Irina was obviously upset, her thoughts occupied elsewhere. When Michael probed a little, she blurted out that a good friend had attempted suicide by drug overdose the night before.

"All his friends, each of us, have been taking turns sitting by his bed in the hospital, holding his hand through the day. But the doctors still are not sure that he will live. He doesn't yet seem to want to live."

Michael listened to her story and tried to comfort her, but he was also curious about the prevalence of suicide, especially among the young, in the Soviet Union.

"I read a newspaper report that said more than sixty thousand people each year commit suicide in the USSR." Irina wasn't shocked. "Why does this happen?" he asked.

"The pain of ordinary life is so great. Everything is a problem here. There are shortages and lines every day. . . . Nothing works and life is hard. Sure, we are not starving, but food is rationed. Children are dying. There's more crime now. . . . Then, when you hear about better conditions in other countries, it causes you to despair about your own. Five years of hopefulness under Gorbachev, but still no solution. . . . "

"Has your faith in God brought some hope?" Michael asked.

"As I said before, I was working with the Slavic Gospel Mission, which brought many Bibles to my country several years ago. They offered me a Bible—a beautiful one. I wanted to take it but I didn't dare."

"Why?"

"I was afraid of what would happen to me. So I politely refused without being able to explain. It wasn't until just two years ago that it was safe to possess a Bible. Now I have one, given to me by the YMCA. And I've read the whole book. It's beautiful!"

At this point, the other interpreter, twenty-three-year-old Nadia Vyalkova, who had been with Sharon and her group at

Thanksgiving, joined the discussion. "When I was a child, I wanted to go to church and meet God, but my teacher said that only ignorant people believed in God, that the Bible was full of fairy tales, and religion only hurt people. She said the church and priests were no good.

"But my grandmummy, the religious one in our family, told me not to listen to my teachers because, she said, after they taught me there was no God, they would go straight to church to pray."

"What would have happened if you had gone to church to find out for yourself?" Michael asked.

"I would have been punished, I'm sure."

"When did it become OK for young people to attend church?"

"Just a couple of years ago. Now almost all my classmates go to church and light a candle, especially when they have exams coming up. Mostly it's just fashionable to go to church. I think they believe that something spiritual exists, but they don't understand it."

Irina spoke up again. "Newspaper articles are talking about this a lot now. Some are written by priests. They say that people are rushing back into the churches, but for reasons that aren't really spiritual. They say that we've lost our soul and our cultural identity in the past seventy years of Sovietization. Now people are going to church to restore faith in our cultural roots as much as to find a faith in God.

"For many people I know, lighting a candle in church is part of our search for our very soul, for the meaning of life and the source of real hope."

Rebecca thought it was delightfully odd that this passionate discussion about personal faith was taking place in an intensely secular setting while, all around, people were celebrating the sanitized state version of the winter holiday.

As the evening progressed and the vodka and the mandates to drink "to the bottom" flowed, the toasts grew more embroidered and sentimental. Michael rose to toast the newly emerging leadership in both countries: "May the new year bring changes of association to the Byelorussian Children's Fund. May it no longer be a fund named after Vladimir Lenin; may it be a fund named after Vladimir Lipsky."

The Byelorussians roared at his boldness and the notion that such a change could even be contemplated. No one could have guessed that by the end of summer, the course of events would no longer make that notion seem particularly bold.

Paul pulled out a bottle of California chardonnay bearing the label of St. George and took the opportunity to tell the story of his and Michael's meeting with Alexander Vasiliev at the United Nations mission under the bronze statue of St. George slaying the "dragon" of nuclear missiles. He then raised a toast to the dragon-slayer, a favored saint who is honored on November 25 on the Russian Orthodox calendar.

"This night is a special night," Lipsky rose to announce. "The last minute of the year of 1990 will be sixty-one seconds long so that astronomers can adjust the atomic clock. So the world tonight is being favored with an additional second of life. We must use it with joy and hope."

At midnight, everyone was standing with glasses in hand, greeting each other with three kisses—two on one cheek and one on the other in true Russian fashion. After the toasts of personal and cultural friendship, Paul raised his glass "to the hope that as midnight moves westward across the time zones of Europe and America, people of good will share their desires for peace and acknowledge the God who moves within us all."

Finally, Lipsky offered a Byelorussian axiom as a benediction: "Remember, however you spend the first day of the new year, so shall the rest of the year be."

6

"WE NEED A SAFE PLACE TO LIVE"

THE FIRST DAY of the new year didn't begin until noon, when both the Americans and their hosts managed to pull themselves out of bed in time for brunch. They had to hurry; there was more celebrating to be done at the Minsk Sports Palace, where Byelorussian gymnasts Olga Korbut and Nellie Kim had trained before the 1976 Olympics brought them international recognition.

At least five thousand children and their parents were gathered in the imposing Sports Palace for the New Year's performance, which was a cross between a high school pep rally and a local talent show. The children sat remarkably still, hanging on every word spoken by the master of ceremonies, who led them in holiday songs and loud clapping.

The hall was draped with banners depicting winter forest scenes. Three colored spotlights—red, green, and blue—danced around the auditorium. A dozen more sat on the stage, illuminating a spectacular winter set hung with elaborate tapestries featuring the gnomes, elves, and mythical beasts of ancient Slavic legends.

The show opened with spirited sing-alongs, dancing clowns, and the dramatic arrival of *Dyed Moroz* and *Snegochka* (accompanied by a crocodile, a cat, and a fox from the children's story Pinocchio). After each character talked with the children and entertained them with songs and dances, a full dramatic production was launched.

Although the Americans could understand nothing that was said, they were able to follow the plot: Grandfather Frost announced the arrival of the New Year and a fir tree was cut down, brought into the house, and decorated with colored paper and a magnificent red star on the top. Suddenly, the cat and the fox appeared and stole the star from the tree.

Grandfather Frost sent for the police, who captured the robbers. The hero who found the star and saved the day, of course, was the crocodile, a favorite storybook character in Byelorussia. He dispatched a snowflake (a young woman costumed all in pink) to climb a tightrope and retrieve the star that had been tossed into the sky.

Ultimately, the cat and the fox were brought to a point of repentance and rehabilitation. The star was restored to the top of the New Year's tree, to the delight of all the children, who were beside themselves in anticipation of what would come next—the opening of their New Year's presents.

As the Americans strolled out of the Sports Palace, Michael was amused. "Look how clever the Soviets have been. They outlawed Christmas as a religious holiday, but at the same time they preserved nearly all its traditional elements in a winter holiday celebration. Saint Nicholas is transformed into Grandfather Frost, and the red Soviet star replaces the bright star of Bethlehem. When someone dares to steal the star from the New Year's tree, he is captured and rehabilitated by who else but the state.

"It's not going to be easy to restore the true meaning of Christmas. It really was a stroke of genius, Paul, to bring the nativity crèches along as visuals to help in telling the Christmas story. Let's use the toys we brought to convey the idea that gift-giving is an extension of the tradition of the magi."

FROM THE SPORTS PALACE, the Americans were taken to a secondary school of art where children with exceptional talent and aptitude, some as young as six, learned painting, drawing, graphic design, ceramics, weaving, sculpture, traditional Byelorussian straw crafts, and even model boat-building.

For the rest of the afternoon, the Americans were treated to demonstrations of skill and artistry by the students at the school. Clay figurines representing characters from fairy tales and Slavic legends were favorites with the visitors. They were offered samples in hopes that orders for more might be generated in America, and perhaps a small business could be started in the spirit of *perestroika*. By the end of the tour, Paul had commissioned five thousand straw star ornaments to be given to his Citihope constituency the following Christmas. It was clear that the local hosts didn't want to miss a chance for an American business connection—even a remote one through a church-based organization like Citihope.

LIPSKY HAD STILL another surprise in store for his American guests. The bus stopped at the edge of the only neighborhood in Minsk that had survived, at least partially, the German occupation of the Great Patriotic War. A tiny cluster of renovated shops, restaurants, and apartments on the bank of the Svisloch River, the Old City dates back to the twelfth century. Tucked into one of the historic buildings was the atelier of Alexei Koozmitz.

A dear friend of Lipsky, Koozmitz's studio was upstairs overlooking the river, and it smelled appropriately enough of oils and newsprint. The loft was lined with canvasses, most of which depicted variations on the theme of a full-breasted and unadorned madonna and child.

"These bold paintings have been so controversial," Lipsky said, "that they were illegal in this country until just a few years ago. In fact, Alexei himself was imprisoned for awhile because he practiced his 'criminal' art form. Now, of course, he has become a celebrity and we must find a sponsor who will spread the word of his work in the West."

Which is why, Michael mused, *the now famous and (surely) well-connected priest from America, Paul Moore, and his delegation have been invited to see his bare-breasted masterpieces.*

Alexei spread a table of sausages, black bread, fish, champagne, and vodka as appetizers, and he told the Americans how to savor good Russian black bread: "First, you must swallow your glass of vodka to the bottom. Then, immediately, you put a piece of bread against your nostrils and inhale deeply. The essence of the bread will never smell better than in that sweet, sensual instant of clean and clear sinuses."

Well, maybe, thought Michael, who suffered from a chronic sinus condition.

Then Alexei made a dramatic presentation of two of his paintings, one to the Moores and one to the Laird-Christensens, for them to take back to America. Paul felt deeply honored; Michael felt uncomfortably obligated. The evening turned to night as every painting in the studio was examined and praised. Some were striking, including one depicting the hell of Chernobyl, but most were different renditions of the motherhood and childbearing theme.

After a series of toasts to the artist's work and sentimental discussions of the possible meanings behind the outlawed madonnas, the relaxed mood got the better of an expansive Paul, and he promised, "You must come to America. We must sponsor an exhibit. Americans will love your work."

Rebecca—who was less than impressed with what she had seen—nearly choked, and Michael gasped. He rescued his friend with a nudge and a forceful whisper: "Please don't make any more rash promises, at least until tomorrow."

With that prompting, Paul recovered to backpedal his promises, and ended the visit with a simple toast to a new and enduring friendship that left the future in God's hands. However, Alexei did get his trip to New York later in the year, but the result fell short of his expectations. Not one of his masterpieces sold.

The evening was completed with dinner at a lovely restaurant on the riverfront in Old Minsk. Although the dimly lit cafe was loud and smoke-filled, the Americans enjoyed the fare and joined the local celebrants on the crowded floor to dance to the lively music of a live New Year's band.

❊ ❊ ❊

PAUL JR. had not been able to wait for the string of official hospital visits. He and one of the interpreters had dropped out of the planned agenda and slipped away to see Natasha Ptushko at the Oncological Hospital in Minsk. His mother had talked to oncologists in New York about treating Natasha's non-Hogkins lymphoma with a bone-marrow transplant in the United States. He wanted to tell her that there was reason to hope and to encourage her to build up her strength for the trip to New York that would be arranged soon.

But PJ was too shocked by what he saw.

Her last letter to him had been upbeat: "I am feeling stronger now. I've gained some weight, which pleases my mother and my doctor, and, for the first time in four months, I have taken some steps on my own. I'm sure I will be even better when you come for the winter holidays."

But Natasha had obviously had a serious setback since then. Despite the remarkable clarity of her eyes, she was too weak to stand. Speech was a great effort, and she conveyed her joy at seeing PJ again mostly through squeezing his hand—which drained her of almost all the energy she could muster.

PJ spent every free moment he had at her bedside, talking to her, playing tape-recorded music, stroking her hands, and even holding her frail body in his arms as he sat on her bed.

Even when she couldn't respond, even when she seemed to be sleeping, PJ told her the Bible stories that he had learned as a child—Noah's ark, Jonah and the whale, and the miracle of the crucifixion and resurrection of Jesus. He talked to her even when there was no interpreter on hand to translate; the words themselves were less important than the message of caring and love he was conveying with his presence.

On one visit, her doctor talked with PJ's father just outside her room. "Her heart and lungs are very bad. I think that Natasha must live, but as a doctor, I think she'll feel better if she didn't. Nobody can help her now. But she still lives with hope. When hope stops, she will not live."

When PJ left her he was agitated, unconvinced that the doctors were doing all they could.

"As soon as we get back, we've got to find a way to get her to the United States for the transplant surgery. She won't survive if she has to wait much longer."

"I tried to find the words to tell PJ that it was already much too late," Paul wrote in his journal, *"but I couldn't. How does a father say those words to his son about someone who means as much to him as Natasha means to PJ? Instead I promised to press the [Byelorussian Children's] Fund to move her to the top of the list of child advocacy cases they were promoting."*

However, the doctors at the Oncological Center were already preparing for Natasha's mother Alla to take her home; there was nothing more that could be done at the hospital.

A week after the Americans returned home, the BCF responded to Paul's request to speed up the paperwork so Natasha could be flown to the United States for emergency surgery. It was the response that Paul expected: *"Natasha Ptushko's mother took her back home from the hospital some days ago. At home, Natasha's state became much worse. Now the girl is again at the hospital and her state is very, very hard. Her transportation to New York is now out of the question. The doctors say she will not stand even the distance to Moscow. We had better wait and hope for her condition to get better."*

❊ ❊ ❊

AFTER DELIVERING methotrexate, antiobiotics, and medical supplies to Dr. Olga Aleinikova at the Children's Hematological Center, and pocketing her wish list for the next shipment, the Moores and the Laird-Christensens split into two groups to offer Christmas wishes and pass out gifts to the children in the wards.

With thoughts of her baby daughter Rachel in mind, Rebecca knelt down next to a lively, bald-headed five-year-old in a red and purple plaid shirt, blue wool pants, and light blue sneakers. Vitya was feeling quite chatty, and he responded enthusiastically to Rebecca's interest. He said he had a brother named Sergei who was sixteen months old, that he lived in Borisov, eighty kilometers from Minsk, and that he had been at the Hematological Center for four months. His mother had been living at the hospital with him the whole time.

"But what about your baby brother? Who looks after him?" Rebecca asked.

"My papa. One week papa works and the next week he spends with Sergei. *Babushka* takes care of him the rest of the time. But I wish papa would not work and would come and be with me."

Vitya talked about his dog who was just a puppy, his cat named Timothy, and his fish. "I'm worried that my papa might forget to take care of my animals and they might be hungry. I want to go home right away so I can feed them and play with my friends. And my mama can take care of Sergei.

"Summer is coming, you know. I can feel it on the tip of my nose."

Rebecca mused at his comment about summer, since it was just the first week in January. Apparently the longing for summer was imbedded in the Russian soul at a very early age, a longing that was reinforced by a lifetime of cold and bitter Soviet winters. Earlier on the trip, one of the hosts had described the two seasons in Byelorussia this way: "We have two months of summer and ten months of waiting for summer!"

Sharon came up to the two of them and asked Vitya what he wanted to be when he grew up. "Perhaps a politician or a comedian. But probably a butcher because I like to eat meat. If I was a butcher, there would be no shortages of meat. I would see to it."

As Rebecca and Sharon left, Vitya wished them good health and long life. "You, too, Vitya. You, too," Sharon replied.

THE AMERICANS GOT A TASTE of what real coordination of efforts might achieve when they met with the Byelorussian minister of health and his deputy the next morning.

The group gathered at the Ministry of Health building, around another blond wood conference table and under the ever-watchful gaze of Papa Lenin. Vasily Karakov outlined the role of his ministry in dealing with the ongoing health crisis of Chernobyl.

"It's our responsibility to hold the morale and the soul of the people in trust. Their peace of mind has been broken and they are anxious. They need to be reassured. But we must also maintain the physical health of the people, especially the children, whose thyroid glands, hearts, kidneys, and livers are being significantly affected.

"We are also forced to do this important work with a real shortage of doctors. Doctors themselves are afraid and move their families out of the contaminated regions as soon as they can.

"So, without enough doctors, medicine, or medical equipment, we are monitoring 173,000 people who are under strict and constant medical supervision. That figure includes 37,000 children. Of course, all of them are just a portion of the 2.2 million, including 800,000 children, who live at risk in the contaminated areas."

To begin to deal with this overwhelming task, the Ministry of Health formed a joint venture with the Byelorussian Children's Fund to open the Anti-Chernobyl Diagnostic Center in a clinic that once served the Communist Party *apparatchik* in Minsk.

"Alber Likanov, a famous writer and the chairman of the Soviet Children's Fund, has donated a million rubles for this radiological center for children in order to call attention to the problems of the children of Chernobyl," Karakov said. "Now it is up to us to supply it with the best equipment and medicine to serve as many children as possible."

He admitted that the Anti-Chernobyl Center, which had been dedicated a month earlier, was far more modest than the USSR Chernobyl Center in Kiev, but he hoped to find sponsors for each component in the plan for the center. Karakov was hoping that Citihope would choose a piece of the master plan to underwrite, and the Moores discussed the possibility of taking on the responsibility of obtaining specialist training for Byelorussian doctors in the United States.

Cornell University-New York Hospital had already agreed to provide the training, as long as Citihope was responsible for room and board while the doctors were in New York. The BCF and the health ministry would take care of transportation.

Another proposal called for Citihope to consider sponsoring a mobile diagnostic unit to help the seriously overburdened district hospitals in the Gomelskaya region. "Not long ago a German van arrived packed with food," Karakov said. "We were very grateful for the food, but what we really needed was the van itself. We are not as hungry as you think; we can get food if we work at it. But we cannot get a van or the medical equipment to stock it.

"And, of course, we will put a plaque at the center to honor those who help."

Michael's mind was racing ahead a bit. *How much equipment,* he thought to himself, *would Citihope need to provide for the ministry to name the center for Jesus Christ? Or the Good Samaritan?*

Paul explained how Citihope connected resources to needs by telling the stories of people in need on their radio broadcasts. Karakov was enthusiastic.

"Yes, yes! We need the solidarity of people of good will around the world.

"We are trying to cure the body; you can cure the soul. We will succeed when doctors and priests work togther."

Perhaps, but Michael was beginning to get a sense of the depth of the problems involved. Doctors and priests might be able to pull it off, but all too often politics and bureaucracy tended to get in the way of their efforts.

❊ ❊ ❊

BUOYED BY THE HOSPITALITY they had enjoyed in Minsk, the Moores and the Laird-Christensens headed into the contaminated Gomelskaya region of Byelorussia, accompanied by writer Beth Lueders and photographer Greg Schneider of *Worldwide Challenge* magazine.

Gomel, known in Polish as Polesje, the Country of Forests, was once the place where Byelorussian parents sent their children to camp, to live close to nature and away from the contaminants of urban society. Now Gomelskaya is the contaminated region, the area of Byelorussia that lay directly under the cloud that rained radioactivity for days and days after the explosion and fire in Chernobyl's reactor No. 4.

Two of the medical centers most in need of donations of drugs and supplies were in the small cities of Narovlia and Gomel in the evacuation zone, just outside the thirty-kilometer total exclusion zone. Residents who wanted to leave Gomelskaya *oblast* had been promised relocation to noncontaminated regions of Byelorussia, but the process was agonizingly slow in the crippled economy of the Soviet Union. People waited, in fear for the health of their children, as weeks and months passed without a single family being relocated.

In the brittle cold of January, background radiation was usually at its lowest levels. When the van carrying the Citihope delegation pulled into downtown Gomel, a digital clock displayed the time, temperature, and the current level of radiation measured in millirads per hour. At that moment, it registered thirty—which was three times the background radiation of most American cities. However, the group was shown photographs of the same sign taken during summer months when the millirad count exceeded one hundred, and the interpreter assured Michael that that extreme level—considered dangerous to human life and therefore rendering the city uninhabitable—was often sustained for several days at a time.

"I learned a valuable lesson on the trip to Gomel," Michael wrote in his journal. *"Our willingness to enter contaminated regions expressed to our hosts a compassion more authentic than words.*

"It didn't seem like such a big deal to us. Experts had assured us that a short stay of a day or two in the contaminated region would not be harmful; the danger was in exposure over time. We were fully aware that we would be leaving in a few days but our hosts, of course, could not. The privilege of our position was unmistakable—and humbling.

"But to our hosts, our eagerness to share their experience, however briefly, spoke volumes."

Uninhabitable or not, people lived—and sometimes died—in the Gomel region, where Dr. Irina Kroyter, the head physician at the Infant Orphanage, had her own hand-held geiger counter—an illegal possession that at that time could have invited criminal prosecution. A strong, imposing woman, she didn't appear at all concerned about the legalities of the jeopardized life they all lived in Gomel.

"Everything is contaminated; everyone is in trouble," she replied, adding that the orphanage itself was located in a "hot spot" and should be relocated immediately. "But it's not likely," she said with a shrug. "The radiation is less inside than outside and less in winter than in summer."

And then began the tour that introduced the seven Americans to a horror they could not have imagined. Crib after crib held children up to the age of four, born since the catastrophe in 1986, with the most dreadful and shocking physical abnormalities, often coupled with such mental conditions as Down's syndrome and hydrocephalus. These were children without hands or feet, without openings for their eyes, with heads twice as large as they should be.

"Did Chernobyl do this?" the Americans asked.

"Who knows for sure?" Dr. Kroyter responded. "All we know is that it is much worse now and the people certainly believe that the accident is to blame.

"Alcoholic parents are known to give birth to children with such defects. Drugs, alcohol, and radiation are all factors. However, the incidence of these congenital abnormalities in Gomel has doubled since Chernobyl, and Down's syndrome cases have nearly tripled. Of course, corrective surgical procedures are not available."

Michael was concerned about the cause of these atrocities, and research revealed that this part of Byelorussia was already seriously threatened by ordinary industrial pollutants before the explosion at Chernobyl. But it was also located right in the middle of the radioactive swath of cesium 137 fallout that the accident had created. The combination of effects from ordinary and extraordinary environmental hazards may have created a most potent, gene-altering mix.

Many of the one hundred and fifty infants at the orphanage, in fact, had parents who had rejected them. "Many were sent to us," Dr. Kroyter said, "directly from the maternity hospital after their mothers refused to look at them.

"Some have parents who were not able to care for them, and some truly have no surviving parents. It's likely that these children will always be wards of the state, since there is almost no chance that they will be adopted."

Dr. Kroyter added that the number of children who are wards of the state is increasing, "not because they have no parents, but because many parents can no longer raise their own children.

"Child abandonment has become a serious social issue in this country. It is so difficult for people to raise children today with food—especially milk—in such short supply, that many parents just give up. The state has even begun an education campaign to urge parents not to abandon their children, to keep them and give them a home."

Some compassionate adoptions did occur, including one child the group met who was being adopted by an American family. Dr. Kroyter said she had no objection to such a practice as long as the request came through the Ministry of Health. This possibility encouraged the Moores, who had entertained thoughts of adopting a Byelorussian child.

The orphanage itself was remarkably clean and comfortable, with playrooms and play equipment, a kitchen, and quarters for the five doctors, twenty-five nurses, and fifty-seven aides on the staff in 'round-the-clock shifts.

"What happens to the children when they reach age four?" Michael asked.

"They are transferred to another orphanage, or adopted," Dr. Kroyter said. "Or perhaps their parents will be able to take them back."

"And when they become adults?"

"Invalids who cannot be cured are taken to the institution."

Photographer Greg Schneider, who had been recording the images of these terribly malformed infants, asked if the children were happy. "Of course not," Dr. Kroyter snapped. "You see for yourself; they have congenital abnormalities. It's dreadful!"

Greg was trying to make a point about the power of love to affect the quality of children's lives, even the lives of the children in her care. "If these children cannot be happy, why should they continue to live?"

Confused by the question, she replied that the state cannot terminate the life of an already-born child.

"That's not what I mean. I want to know if you believe that the power of love can make life—even these lives—worth living."

He told her that he believed even the most dreadfully malformed children—like two-year-old Andrei, whom she held in her arms and who had the grotesque features of advanced hydrocephalus—were capable of happiness if they were loved.

Dr. Kroyter smiled but she didn't respond. It may have been a novel notion to this good doctor, who obviously loved the children in her care, that they might be able to find happiness in their compromised existence.

❊ ❊ ❊

THE REST OF THE DAY was spent at the collective farm Vu Pokalyubichi, where food was still produced by several hundred workers who lived and worked amid background radiation that exceeded acceptable standards in any other part of the industrialized world.

The chairman of the collective, Vitaly Zheleznyak, was a friendly bear of a man with gold-capped teeth and a broad smile. He had held his position for sixteen years, after taking over when the former chairman suffered a heart attack. At the time he had been chairman of all the collective farms in the region, and his tour at Vu Pokalyubichi was supposed to be temporary. But he ended up staying in this place, which he clearly loved.

The collective had been founded in 1929 as a communist experiment when Stalin's forced collectivization took the land from private hands and made farming a state enterprise. The visitors watched a documentary film, "The People of Our Land," that celebrated the values and the successes of hard-working farm people who labored to feed the vast Soviet empire.

Michael couldn't help feeling that this model collective no doubt had been a primary stop on the tour of communist success stories during the sixties and seventies, but the film was just so much empty propaganda now that the world was aware that food shortages in the Soviet Union were reaching critical proportions.

Vitaly led the Americans around the three-thousand-acre farm, showing off land devoted to the production of grain, potatoes, wheat, and a factory that roasted Colombian-grown coffee beans. As the tour proceeded, the Americans kept waiting for the chairman to mention how the explosion at Chernobyl had affected the collective, but nothing was said. Finally Michael asked the question on all of their minds: "Mr. Chairman, is the food grown here contaminated?"

"The radiation level of the food is acceptable," he said without blinking. "There are research scientists here from the institute conducting experiments, analyzing the food and comparing what is grown here to produce from other regions. The milk and beef are measured for radiation. The level generally is not high enough to throw it away. Of course, what is not acceptable is discarded.

"We found that corn, for example, does not absorb radiation and is not contaminated. We feed our cows corn and hay. We wash them inside and out after the slaughter and all are monitored. They are found not to be contaminated. Other cows who grazed near Chernobyl did register high in radiation. They were bloated and their hair stood on end like porcupines. They were slaughtered and buried."

Michael persisted in his line of questioning. "Do you send beef and food grown here to other regions of the Soviet Union?"

"Frankly speaking," Vitaly said, "nothing we grow here is wanted in other regions. Still, we manage to ship some out."

In fact, dairy and meat products from the contaminated regions that exceeded the established radiation limits were regularly distributed throughout the Soviet Union to minimize the chance that one segment of the population would end up with a dangerous concentration of contaminated food. This was part of a national policy called "universalizing the exposure."

The chairman's jowls quivered and his bushy brows wrinkled when he admitted that the industry of his sixty-year-old collective was useful to no one. "Ideally, it would be better not to plant anything here. . . but taking into consideration our very poor economic situation, we must grow what we can with precautions and evaluations."

"What about your children?"

"We don't feed our children the food we grow. . . . We buy uncontaminated food from other regions."

"Are any of your children sick?"

"Sixty-eight of our children have enlarged thyroids and have experienced headaches and nosebleeds—the first signs of radiation sickness. Twenty-four are ill and have gone to Minsk for treatment."

With obvious pain, he described elaborate efforts that were under way to get their children off the farm and out of the contaminated region for the summer so that sunshine, exercise, and good food might strengthen their jeopardized immune systems. "They go to central Russia or Czechoslovakia."

It must have been particularly galling, Michael thought, *for the chairman of one of the few truly productive collective farms left in the Soviet Union to have to admit that the success that he helped build at Vu Pokalyubichi had been wiped out by the failure of some alien technology across the border in the Ukraine in 1986.*

As the group huddled in the bitter January cold, several mothers from the farm gathered around Sharon Moore to plead for help for their children.

"Our children eat bad food at school and at home because we do not have enough clean products. Only bad vegetables and fruit are at the market, and the prices are very high—."

A bewildered father interrupted, saying, "We can't do anything because we are simple people. We knock at every door and can't get help. Nobody pays any attention to the simple people in this collective farm."

"We want help with vitamins and special medicines for our children," pleaded yet another crying mother. "Our children have headaches and every day they have blood from their noses. We have no medicine at all."

The barrel-chested chairman stood to the side and made no attempt to stop the outpouring. A few years earlier, he would not have allowed such a display. His pride and the pride of the collective would have prevented it. But now, nearly five years after the catastrophe, it was too late for pride. If begging would bring help for their children, then they would beg.

Sharon was distraught. "We came here to talk about God's love, but you can't talk about God's love when people are starving. They can't hear over the rumble of their own stomachs."

Eager to get inside once again, the Americans were trooped through each of the schools of the collective—kindergarten through junior high—and by now Paul and Rebecca had their magi routine down pat. This time, though, Paul was inspired. At the elementary school, after the dance around the New Year's tree and the program of songs and music, Paul decided to tell a new version of the Christmas story. With thirty children sitting on the floor, surrounded by twenty or more adults, all eyes were upon this imposing figure in a cleric's collar and fur hat.

In his deep, mesmerizing voice, Paul began:

> *A long time ago in the village of Vu Pokalyubichi there was a collective farm. It was winter and there was snow on the ground. All the children had homes to go to and parents to keep them safe and warm. Life was difficult but good.*
>
> *One day, visitors came to the village. Mary and Joseph were their names. (Paul pointed to the figures in the nativity scene the group had brought as a gift to the children.) Mary and Joseph were from another country. Mary was pregnant and about to give birth.*
>
> *They inquired at the village about a hospital. There was no hospital in the village. They asked if anyone had a place to stay for the night. A room was found near the animals in the barn. It was the only place available on this Christmas Eve.*
>
> *Suddenly there were angels in the sky! (Here Paul picked up the angel from the crèche and flew it around the room as the children giggled with delight.) The angels spoke to the farmers and told them the good news—that on this night in their own village would be born a child who would become king of all the world. They said the village was honored to be chosen as the place of his birth and told them to go right away to see the baby. (The children were awestruck and wanted to touch the figure of baby Jesus that Paul cupped in his hands for them. The adults were just as engaged.)*
>
> *Then came the magi to the farm with their gifts—uncontaminated food, toys, and medicine—for baby Jesus. And all the children on the collective farm sang for joy and danced around the tree that was especially decorated for the occasion.*
>
> *The children from the farm village of Vu Pokalyubichi were not forgotten by the God of the universe, who so loved the world and the children of the village that he sent his only son to be born in a barn in their midst.*

Then Paul invited the children to repeat after him in English: "Hosanna to God in the highest. Peace on earth, good will to all! Merry Christmas!"

Once again, Paul asked the children if they had ever heard that story before—and not one raised a hand. "Now that they've heard it, and participated in it," he said to Rebecca, "perhaps they do know and understand the real meaning of Christmas."

✳ ✳ ✳

AT THE END OF THE DAY, as always, there was ample food and hospitality for the guests, even if—as the Americans suspected—the hosts would later do without to provide it. Dinner with one family on the collective provided the Americans with the opportunity for some citizen-to-citizen diplomacy.

Their hosts were Vasya and Yelena Yerstratenko, and it was Yelena's fortieth birthday. A large table was set in the tiny living/dining room, and the seven Americans, the chairman, the Yerstratenkos, and their two daughters—eleven-year-old Olga and seventeen-year-old Inna—crowded into places around it.

After the conversations of the day, the Americans were leery of eating the contaminated food. The chairman sensed their apprehension.

"Perhaps you are worried about eating this food?" Vitaly asked. "You must not worry. The radiation is of an acceptable level. There's no danger."

Not wanting to offend their hosts—and being hungry, the Americans ate. The mushrooms were particularly good, and Yelena explained how many varieties of mushrooms were grown in the forests of the region. Of course, they too had to pass radiation inspection.

As the vodka was poured for toasts, Vitaly, the chairman, told them, "Your own Dr. (Robert) Gale (of UCLA) said that a glass of vodka a day washes the radiation away. The vodka absorbs the radiation. A piece of bread absorbs the vodka, and water washes down the bread."

Michael leaned over to Rebecca and whispered, "He's joking, isn't he? Gale would never have said that. He must be joking."

As the evening wore on, the conversation turned melancholy. This time it was Paul who had brought up the dark subject of Chernobyl.

"Do you know what this word means?" Vitaly asked.

"Tell us," Paul encouraged.

"It means 'bitter ground.' Our ancestors who named this region Chernobyl were being prophetic."

Vitaly talked about the day after the explosion. "It was a beautiful, sunny, wonderful day. About four in the afternoon, my grandson and I were walking, enjoying life. . . . We saw a black cloud and it started to rain. Of course, we didn't understand that it was acid rain."

Vasya added to the story: "I too thought it was simple rain. Then it became very dark and there was smoke. . . . Two million people were exposed, and we cannot all survive. We have so many sick children.

"The world must not play with nuclear energy," Vasya said, turning from melancholy to anger. "We need a safe place to live. . . . "

Did they want to move away? "I love my motherland," Yelena answered. "Of course, I can go far away from this place of bitterness, but I don't want to abandon the place of my birth. I'm not alone in this. We all feel this way, so we stay and grow our grain.

"We cannot deal with our problems alone. We need help from other countries. Your love and kindness have helped us understand that we are not alone in our grief and sorrow."

At the close of the evening, Michael offered a pledge: "We cannot control what our government leaders will do. We can only promise among ourselves, friend to friend, to help each other. To help the children, to work for peace, to demand that nuclear weapons never be used."

"Yes," said Vasya as a final toast. "Let us at least agree not to blow each other up!"

"*Absolutva!*"

7

"OUR LIVES WILL BE DIFFERENT FROM THIS DAY"

THE THREE-DAY JOURNEY into the contaminated Go-melskaya region ended in time for the Americans to return to Minsk for the observation of *Rozhestvo* on January 7, with elements of both the Russian Orthodox and American Protestant Christmas traditions included.

The Americans were not prepared for the cultural feast that awaited them at the Pioneer Palace in Minsk. For starters, Michael had effectively lowered everyone's expectations.

"Remember that the Pioneer youth movement in the Soviet Union wasn't just a benign group of Boy and Girl Scouts," he said. "The Pioneers were started by Josef Stalin as a vehicle for indoctrinating even the youngest children with the communist ideology. The clubs were intended to prepare them for party membership."

So the Americans were properly guarded against the political onslaught to come when they arrived at the hall.

Yet what they encountered there couldn't have been less political or more moving and delightful—and they were left reeling from the experience. Amid flowers and much pomp and ceremony, the group was ushered to special seats for a cultural tour of Byelorussia conducted by incredibly talented children.

The first group, called "Peace Child," featured nearly a hundred teenagers who sang with remarkable spirit and mature harmony. They had recorded a cassette of songs and were in great demand throughout Europe.

Already impressed by "Peace Child," the Americans were overwhelmed by the next act. "Zorachka," which means "little star," included forty children between the ages of 8 and 15 who performed folk songs and dances in traditional costumes. The purity of their voices and the precision of their dancing was testimony to the discipline of their after-school program at the Pioneer Palace.

Paul and Sharon were totally captivated by Zorachka. Always looking for effective ways to carry the message of the children of Chernobyl to the American public, they had separately come to the same conclusion: Citihope must respond to the BCF's request to send Zorachka to the United States for a tour to benefit the medical relief programs in Byelorussia.

However, the delights were not over. The evening ended with a symphony performance of secular and sacred classics by young musicians, later joined by an adult chorale, that transfixed the Americans. Michael thought, *How fitting that Christmas Eve in Byelorussia would be like this!*

Their spirits were soaring as they bundled up and headed into the brutally cold night.

All the performers, of course, were hoping to be invited to tour in America, and they had been led to believe that Paul Moore could move heaven and earth to make such dreams happen. But Paul and Sharon had already decided on Zorachka, and Vladimir Lipsky would not let the Americans go home without first forming a solid plan on when and how the ensemble would get to America. On Christmas Eve, no obstacle seemed too formidable to overcome.

THE REMARKABLE EVENING would end with a traditional American Christmas party in Sharon and Paul's hotel suite, with the Lipskys, Alexander Trukhan and his daughters, interpreters Nadia, Igor, and Viktor, the Moore family, and the Laird-Christensens. Amid the food, drink, carols, and presents, the reason for the celebration was honored as well.

Michael and Rebecca, with Nadia interpreting, led the group through a reading and enactment of Christ's nativity, using the crèche. Everyone was given a part to play as the story was told.

At midnight, Paul offered the Eucharist in the Episcopal tradition from the *Book of Common Prayer*. As he finished, Lipsky and Trukhan were moved to declare their new faith in God, and Paul invited them to partake of the elements of communion. With great reverence, PJ placed a morsel of bread in each cupped hand and Michael passed the cup.

Then Trukhan made a small speech. "It has been the God motive that has brought these Americans to help us and our children. And this is what sets them apart from all the many others we have met and worked with. Now I understand the difference that motivation makes."

Lipsky added his thanks for revealing the meaning of Christmas to them. "I have read some of the Bible, but today you have performed a visual reading of the Bible. And it seems to me that our lives will be different from this day."

Turning to Paul he said, "I have seen the way your son treated the bread and I hope that some day my daughter will treat the bread the same way."

For these American Christians, it was a moment of immense importance that would help sustain them and their Byelorussian partners through many struggles and misunderstandings in the months ahead.

❇ ❇ ❇

CHRISTMAS DAY would not be a private day of family celebration for the Citihope delegation. They had a very public event on their agenda—one that would end up with fifteen minutes' coverage on the evening news in Minsk.

Byelorussia would confer the highest civilian award of the republic on Paul, the priest from America who had marshaled so many resources for the benefit of the children of Chernobyl. It would be the first time the award had been presented to a foreigner.

At different points, each member of the Citihope group had wrestled with why this particular effort seemed to have touched a responsive chord. Other international relief organizations had shipped far more food and medicine than Citihope.

What seemed to make Citihope's efforts unique was Paul and Sharon's determination to accompany each shipment, speak openly about their motivation, and tell those receiving the supplies that the cartons were gifts from people of compassion in America who cared about their children.

Individuals and organizations in the United States took satisfaction in sponsoring a child for fifty dollars. That would purchase one thousand dollars' worth of methotrexate to treat a child with leukemia for a year. The Citihope team members found deep fulfillment in accompanying and supervising the distribution of these shipments directly to the doctors, sitting at the bedsides of the stricken children, and talking to their parents. For those who had made the trips, this was the real reward; they had no doubts that they were the ones who had been blessed.

Michael had written in his journal: *"What a powerful witness it is to explain to the doctors that the medicine we bring is donated by Christians in America who care, to anoint with oil and pray for sick and dying children, to seek to inspire a suffering nation to trust in God. . . .*

"Our visits, ordinary by Western standards, are perceived by the Byelorussians as larger than life, tangible proof that God has not abandoned them."

As they drove to the parliamentary presentation, it occurred to Michael and Rebecca that, once again, political forces were using Citihope to further their own causes. Most of the time it wasn't a problem—as long as the goal of compassionate relief was furthered, along with whatever other agendas others might want to promote.

This time, however, things were clouded a bit. It was no secret that the Popular Front and other democratic forces in Byelorussia were gaining strength, and the Communist party's iron grip on this republic was being pried loose. Could it be that Byelorussia's embattled president, Nikolai I. Dementei, had found a way to bask in the reflected glow of this American priest by bestowing an honorary "diploma" on Paul?

If that was Dementei's goal, Paul was unconcerned. He saw it as an opportunity to offer the message of Christ to people on the verge of great changes in their lives. He read the entire sixty-first chapter of Isaiah, including the passage, "He has sent me to bring good news to the oppressed, to bind up the brokenhearted, to proclaim liberty to the captives and release to the prisoners." Paul also managed to deflect the glory of the moment to the greater glory of God.

Dementei, who would be out of a job nine months later, made some most uncommunist-like statements: "How appropriate that, on our Christmas Day, we welcome Christians who brought love, hope, and healing to our children. And Christ, who himself gave healing to children, is being honored in our nation for the first time in generations."

Pyotr Kravchanka, the foreign minister, added: "It is not simply for your charity that we recognize Citihope, but also for the hope you bring to our people. Do not think this award is only about how many boxes of medicine you bring—as grateful as we are for them—but for the hope and inspiration to our people."

They underscored the point that personal encouragement was better than impersonal charity, that individual medical relief delivered by people of compassion was more meaningful than massive, high-profile airlifts.

Paul later described the award as his "adoption papers." *"I feel I have been adopted by this land, by this nation, by the people of Byelorussia. . . . After 73 years of silence on the nativity of our Lord, Christmas has now been declared an official national holiday. Every Byelorussian gets the day off from work. People are responding to this opportunity to have a holy day."*

The same day, the Byelorussian Children's Fund announced the appointment of Sharon Moore to its board of directors.

Later Michael apologized to Paul for his cynicism. "It finally dawned on me during the ceremony that Citihope's mission had moved beyond connecting resources in America with needs in Byelorussia. You and Sharon have emerged as local celebrities, heroes from America who are able to inspire people to hope again in the midst of despair. At first it was amusing to watch you get treated so royally, and I was the first to challenge you about your celebrity status and remind you that you're human.

"But now it appears that you and Sharon have been chosen for such a time as this—and to whom much is given, much is expected. But are you prepared to be the messiahs of Byelorussia? We all know what happens to messiahs—."

FINALLY, THE REALITY OF CHRISTMAS and the nativity of Jesus Christ would be observed in the majesty of the *Rozhestvo* service at the Russian Orthodox cathedral in Minsk. The church was decorated for the holiday. Boughs of fir were hung everywhere and the huge fir tree in front of the cathedral was strung with brightly lit colored lights.

Worshipers were packed into the sanctuary, and the Americans were squashed against a state television crew that was on hand to record the appearance of the newly ransomed captive— Christmas. When Michael discovered that one of the cameramen could speak English, he introduced himself and asked him what he thought of observing Christmas as a Christian holiday again.

"The state is so desperate, it had to happen sooner or later," he replied. "A drowning man will grasp at any straw."

He told Michael that it had been announced just a week before that four republics—Russia, Ukraine, Byelorussia, and Georgia—would have official holidays on January 7. When Michael informed him that he had been told back in September that it would happen, the man laughed. "Well, it was a surprise to us."

Just before the service began, a bearded, ascetic-looking priest appeared before them. "Follow me, please," he whispered as the crowd pressed in. He made a path through the throng by pushing parishioners aside and led the Americans to places of honor inside the sacred altar rail, in front of the five-tiered iconostasis, which bore the likenesses of the disciples and the saints in brilliant, gold-leaf iconography. Through the open door of the iconostasis—the "window to heaven"—they caught glimpses of the altar, where Paul, Michael, and PJ recognized Philaret—the metropolitan of Minsk himself—at prayer.

From their privileged positions in the flickering half-light of hundreds of candles, they were drawn into the mystical and sensate nature of the Russian Orthodox liturgy. An a cappella choir's anthem of "right praising" spilled over the worshipers, with the bell tones of boy sopranos sounding and then repeating in the domed sanctuary. The hypnotic chanting of the cantor in ancient, ecclesiastical Slavonic, and the congregation's ritual response of *"Gospodi pomily"* ("God have mercy") captivated Michael and Paul. The experience was completely antithetical to the informal, nonliturgical, community-centered worship of their own Nazarene church background, but the two clergymen were swept up by the transcendent nature of the experience.

Michael closed his eyes and allowed the choral chanting to wash over him like a wave, as the parishioners fell to their knees and pressed their foreheads against the cold floor of the sanctuary.

PJ was offended by what he considered to be the excesses of the cathedral's adornments and its liturgical rituals. To him, it was just so much "vain repetition," and he wanted to visit Natasha again. He leaned over to his mother and whispered, "Where is Jesus in all of this? You wouldn't even know it was Christmas Day." Then, as inconspicuously as possible, he backed his way to the side wall of the sanctuary and disappeared into a dark hallway.

Rebecca was surprised at the constant activity that unfolded before them. Only the oldest and most infirm took the seats that lined the walls; all the others stood throughout the service, moving from station to station, crossing themselves from right to left in the Orthodox fashion. Ancient crones, *babushkee*, hovered over the tiered, bronze candle holders. They moved candles around and dropped them into the buckets when they burned too low.

Old women with brightly printed wool scarves covering their heads moved through the congregation with collection plates. Congregants, who came and went throughout the service, dropped kopecks onto the plate and picked up other coins, "making change."

Another full-bearded priest and some lesser celebrants paraded through the ornate sanctuary carrying the distinctive Russian Orthodox cross with the small, slanted crosspiece (representing Christ's footrest) halfway down the stem. The priest was swinging a censer, and the acrid fumes of the incense added another dimension to the service.

After the three-hour liturgy, Paul turned to Michael and said, "This is the perfect time to hand out the rest of the toys on the bus."

Michael nodded and made his way through the crowd to the bus. He hauled the duffle bag and boxes of toys to a spot on the snow next to the huge decorated fir in front of the cathedral doors. He was thinking of himself as a sort of American Santa Claus, grandly presenting toys on Christmas Day to the appreciative youngsters of Minsk. The idea faded fast.

Within seconds after he handed the first child a toy car, he was swarmed. Children and adults piled onto him, pulling toys from the boxes and elbowing him onto the snow. He shouted for

Paul to help him, and the tall priest plowed into the crowd, shoving people aside to get to him. They were in the center of a near riot, and the potential for real violence was growing. Angry and confused, they stuffed the remaining toys back into the bag, zipped it closed, and pushed their way back to the bus.

"We had to stay on the bus for quite a while to catch our breath and our composure," Michael recalled. "From then on, the toys would go to the children in the hospitals—nowhere else."

❖ ❖ ❖

THE NIGHT BEFORE THEY LEFT, the Americans heard the story of still another child in need, named Galina. Her father, Alexis Beliy, was a good friend and benefactor of Vladimir Lipsky. He had joined the group a number of times over the past week. This time he would host an elaborate dinner at his home.

Alexis was a well-to-do building contractor in Minsk who had been living with his family in Gomel when the reactor exploded. He was called upon to help in the construction of the concrete sarcophagus that was intended to encase the damaged reactor and seal in the radioactivity. The state government of Byelorussia also used him to build apartment houses to shelter the "resettlers," evacuees from the total exclusion zone around Chernobyl.

"That was a particularly difficult task," he said. "I was able to obtain few of the building supplies I needed for this immense job. I had to make do with whatever substitutes I could find."

The dinner proceeded along the lines the Americans had come to expect—hours of light banter and cheerful conversation, punctuated by toasts and course after course of exotic foods. Late in the evening the convivial tone turned quite serious, and Alexis shared the numbing story of his fifteen-year-old daughter, who had spent the winter holidays in the hospital with her mother at her bedside.

"My Galina suffers from a serious heart valve defect and she must have valve replacement surgery. I have been told that this surgery is routine and very successful in the West, but here, in the Soviet Union, such an operation is still experimental. No doctor in Minsk will do it for her.

"Tomorrow I travel with my daughter to Kiev for another examination, but I am sure that they cannot perform the surgery there either. Ninety percent of these surgeries here do not succeed," Alexis whispered hoarsely as he blinked back tears.

"So you see I have nowhere else to turn for help but to you." He said he would appeal to all the factories in Minsk for donations of hard currency to make a symbolic deposit for Galina's operation. "I am willing to sell everything I have and give up this wonderful home to save my daughter."

Late that night, Michael wrote in his journal: *I was really moved by the urgent plea of this man. What can we do? Add Galina to our list of advocacy cases? Try to take her home with us? It's all so overwhelming, so depressing. By the end of the evening, we were embracing Alexis, joining him in hoping against hope.*

❖ ❖ ❖

ONE PARTICULARLY DIFFICULT task awaited Sharon before she could leave Minsk. When she had visited Byelorussia at Thanksgiving, a young mother, Claudia Aleksandrovich, had pressed a foil-wrapped package into Sharon's hand and begged her to take the bone-marrow slides in the packet to oncologists in the United States, so that they could review the diagnosis of Hodgkin's disease in her six-year-old son Andrei. Claudia had been told that Andrei's cancer was in remission, but she told Sharon, "He is still sick. I'm sure it is a false diagnosis. Without a true one, he cannot get well."

Unable to deny this mother's appeal, Sharon put the slides in her bag and carried them home with her. Once back in New York, she had taken them to Dr. Jim Bussel, an oncologist at Cornell Medical Center, who told her the slides were virtually useless since they were not fresh. He would need to have new slides delivered within just a few days after they were drawn from Andrei's pelvic bone.

That was the message that Sharon now had to deliver, and Andrei, already small and thin for his age, visibly shuddered when her words were translated for him. Sharon was unaware that she was condemning the youngster to a most painful procedure for a child—since in the Soviet Union, bone marrow was routinely drawn without the benefit of anesthesia.

When this was explained to Sharon, she knelt down and gathered him into her arms. "Andrei, you must do this for me," she said. "You must be a brave soldier and I will become your advocate with the doctors in the United States."

The child nodded gravely to her, but there was no smile and Sharon could feel his small body tremble.

THE POLITICAL SITUATION in the Persian Gulf had not improved while the Citihope delegation was in Byelorussia. They arrived at Sheremetyevo Airport in Moscow with reservations on a Pan Am flight to New York, only to discover that the flight had been canceled.

Actually, they were told, the plane had been commandeered by the United States government to carry more troops to Saudi Arabia. They had been completely out of touch with news from the West for the past ten days. They had hoped that the crisis in the Gulf had been resolved, but it was clear that the United States and Iraq still stood eyeball-to-eyeball five days before the January 15 deadline—and no one had blinked.

It took some negotiating—trivial, really, after the ordeal that had been required to get the six thousand pounds of medical supplies into the Soviet Union when they arrived—but the Americans managed to book seats on an Air Lingus flight the next day that would bring them to New York via Shannon Airport in Ireland.

With the same relief that travelers the world over must feel, they arrived home again—to places where they could speak the language, the food was familiar, and they knew how things worked (or what to do if they didn't).

8

A GATHERING OF ANGELS

THROUGH THE HOLIDAYS, Michelle Carter had carried with her the faces of the children and their mothers in the hospitals of Minsk. Now she would make the time to see what kinds of resources were available nearby on the San Francisco Peninsula.

Early in December, just a few days after her return, she had settled in to write at the computer keyboard in her glass-walled office at the *San Mateo Times*. The process of writing was cathartic for her. She often told people who asked about her impressions from the trip that she needed to sit down at a keyboard before she would really be able to sort things out. A fast and usually terse writer, she was surprised to discover she had enough material for a five-part series for the paper.

After the series was published, readers donated several hundred dollars in response to the article about Dr. Olga Aleinikova and the Children's Hematological Center in Minsk. The donations had a sobering effect on Michelle, who was beginning to realize that she had started something that could not be handled casually. The money would be carefully accounted for, and she was determined that every cent would go for medicine. She was feeling the weight of being entrusted with other people's money.

She also knew that if she hoped to raise any more, she would have to establish nonprofit status for this effort, which she was calling the Children of Chernobyl Project — not very original, she realized, but quite to the point.

After a few phone calls to state and federal agencies, Michelle discovered that the road to "tax deductible" status for an organization was long and arduous, and would eventually require the services of an attorney. With a demanding full-time job, she had no interest in spending time and money on building an organization. She wanted to keep it simple. The last agency she called had the answer: "It doesn't sound to me like you need to go through any of this. There's no need to reinvent the wheel. Why not find a tax-exempt organization with similar goals to serve as fiscal agent for the project?"

What better "tax-exempt organization" for her project than her church, the Congregational Church of Belmont, whose members had laid hands on her head and commissioned her to represent the whole congregation on the November peace delegation?

Michelle had been a lay leader of the church for fifteen years. She had served as moderator of the executive council and on every one of the church boards. Sensitive to the "Congregational" way of doing things, she met with different groups within the church and won enthusiastic approval for the creation of the Children of Chernobyl Project of the Congregational Church of Belmont.

Now donations to the project could be made to the church and covered by its tax-exempt status. This step also allowed vendors of medical supplies who donated goods to take a tax deduction for them. This turned out to be a very important point when the "angels" who began to find resources for the project needed to apply a little persuasion.

Her first angel was Lee Gianfrancesco, the supply nurse at San Mateo County General Hospital. Michelle served on the board of directors of the foundation of County General. After she had shared some stories from her trip at a board meeting, the executive director suggested she contact Lee.

"He's already gathered a large supply of sterile syringes and needles, which were intended to go to children's hospitals in Romania. But I don't think they were ever able to arrange for the shipping. It was far too expensive. Knowing Lee, he probably still has everything tucked away, waiting for just the right cause."

Michelle filed away the thought about shipping and finding a pipeline to Dr. Olga's hospital, and called Lee. He loved the idea. He already had a substantial "stash" of supplies that the hospital could no longer use, but he also knew vendors to contact

Vladimir Lipsky—author, editor, and president of the Byelorussian Children's Fund.

In January 1991 a Red Army soldier stood outside the Orthodox cathedral in Minsk, seeking help for his daughter, who suffered from cancer.

Michelle Carter raising funds at a July 1991 luncheon in San Mateo, where she was the guest speaker.

The first Citihope fact-finding mission to Minsk included (left to right) Paul Moore, Jr., Paul Moore, and Michael J. Christensen.

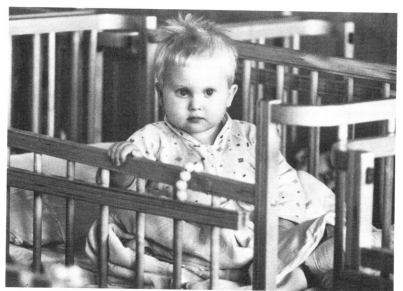

Chernobyl's radiation is also taking its toll on children born years after the explosion.

A caretaker hugs a sad boy in an orphanage in Gomel, in the contaminated region in south-eastern Byelorussia.

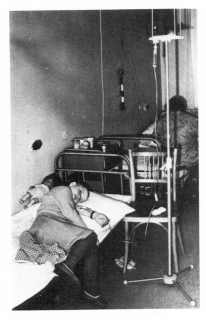

A child with leukemia at the Children's Hematological Center in Minsk. Dedicated physicians cope with staff shortages and inadequate facilities and equipment.

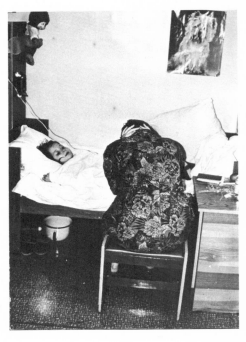

A Byelorussian mother despairs over her hospitalized daughter. Mothers (or grandmothers) almost always stay at children's bedsides. State factories grant paid leave to care for a sick child.

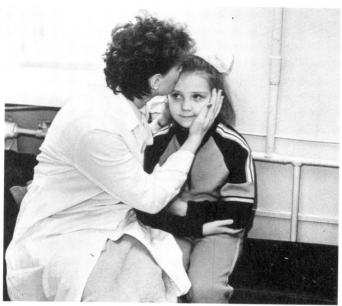

A nursing attendant at the Children's Hematological Center knows that both love and medicine are lifegiving gifts for a child with leukemia.

Dr. Olga Aleinikova, director of the Children's Hematological Center.

Dr. Olga's highest priority for her patients was to obtain methotrexate, an effective drug for treating leukemia that is unavailable in the Soviet Union. One small vial could be purchased in Europe for $200 in hard currency.

Paul Moore, Jr. ("PJ") became close friends with fourteen-year old Natasha Ptushko, a critically ill cancer patient.

Paul Moore, Jr. with children at an orphanage.

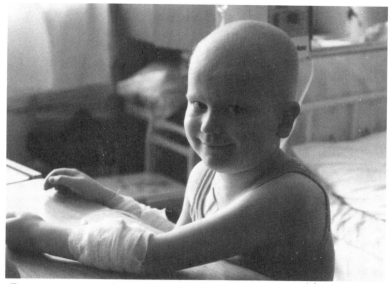

Cancer treatments have resulted in hair loss for a young boy

Children at a Minsk orphanage enjoying toys delivered by a Citihope team along with with food and medicine.

Paul Moore shares a joyful moment with one of the children at an orphanage in Gomel in the contaminated region.

Rebecca Laird-Christensen paraphrased the biblical story of Christmas for Byelorussian children who were hearing it for the very first time.

Paul Moore and Rebecca Laird-Christensen acting out the story of Christmas for Byelorussian children—with the help of a crèche and an interpreter.

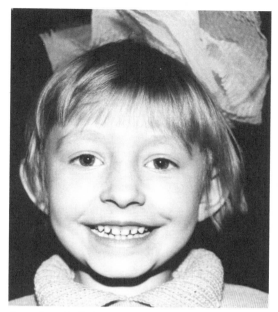

In spite of all that they have suffered, children of Chernobyl have not forgotten how to smile.

Sharon Moore became well-acquainted with Dr. Olga's hematology patients.

Vova Malofienko attended Paul Newman's
Hole-in-the-Wall-Gang Camp and stayed on
in the United States for two years of medical
treatment in Boston.

A reminder about the persistence of radiation from the Boston
observance of the fifth anniversary of the Chernobyl explosion.

A family in the contaminated region of
Byelorussia, where the radiation level measures
150 millirads per hour—fifteen times the normal
level.

A woman from the
contaminated Gomel region
of Byelorussia.

A family whose home was razed by the government
in the total-exclusion "Dead Zone" within thirty
kilometers of Chernobyl.

Sharon Moore inside an abandoned school building in the Dead Zone.

Michelle Carter at a guard station in the Dead Zone, near Chernobyl.

For the first ten days after the Chernobyl explosion, citizens within the Dead Zone were given no warnings that there was any danger; then they were given half an hour to leave their homes and their dreams behind.

Svetlana Lukashuka read about the coup that unseated Mikhail Gorbachev while at camp in the Sierra Nevada Mountains with Byelorussian children. Back home, her husband was demonstrating in the streets of Minsk.

Outdoor activities were favorites for children of Chernobyl at camps in the United States. High levels of radioactivity prevent them from enjoying nature at home.

The memorial at Khatyn, honoring the 186 Byelorussian villages that were burned to the ground by the Nazis during World War II. This father, carrying his dead son, was the lone survivor of the village of Khatyn.

Asa Tribble with farm couple in abandoned village near Chernobyl.

By June 1992, the scale of Citihope's activities had increased dramatically. Instead of hand-carrying supplies to Minsk, sixty-three volunteers would accompany 230 semi-truck trailers of food and supplies to more than three thousand villages, hospitals, orphanages, and institutions.

Paul Moore, Jr., loading sacks of rice during "Operation Hope Express."

Sharon Moore placed an "Operation Hope Express" logo on a military helicopter before accompanying medical supplies to Gomel.

Michael Christensen led a group of sixty Americans to Kurapaty, a recently discovered mass graveyard for victims of Stalin's atrocities.

A worship service held for the doctors and nurses at the Anti-Chernobyl Diagnostic Center in Minsk.

who were likely to be able to help as well. And he shared some advice that would serve her well: "Forget the hospital administrators and chiefs of staff. Go directly to the chief supply nurses at local hospitals. They're the ones who know what's available and how to find it."

Lee also offered something more than advice. "I'm a quilter. I always have two or three quilts going at my house at one time, and I like to give them to causes that I support. Would you like one of them to raffle off for the children of Chernobyl?"

Lee made the offer in such an offhand way that Michelle was quite unprepared for the exquisite, double-size, hand-stitched quilt in shades of teal and burgundy that he handed over a few weeks later. She stored it in a zippered plastic pouch and took it (along with raffle tickets she sold at one dollar a piece or six for five dollars) with her any time she spoke to groups.

In April she displayed the quilt at the annual meeting of the Northern California Conference of the United Church of Christ, held in Asilomar. A bazaar was planned after lunch on the Saturday of the four-day conference, to be set up in the meadow just across from the beach on Monterey Bay. The quilt was an instant hit. Michelle and her husband, Laurie, could not keep up with the press of the crowd trying to buy raffle tickets, so they snared friends passing by to help.

By the end of the weekend—and the end of the raffle—Lee's handmade quilt had raised more than fifteen hundred dollars for the children of Chernobyl.

Lee was right about contacting the supply nurses. Some were already committed to gathering supplies for other causes, but many angels of mercy were ready and willing to help with this one.

It was an education for Michelle that so much was available. How could fiscally responsible hospitals have so much "extra" that could be given away?

"Oh, it's easy to explain," Lee told her. "In California, hospital codes—which are already among the most stringent in the country—are constantly being upgraded. When a code is changed, a hospital could be stuck with a sizable inventory of a style of catheter that could no longer be used in California, but would be perfectly acceptable in Nevada."

"And practically priceless in Minsk," Michelle added.

Vendors sometimes got caught by the changing codes as well. The state of California made a slight alteration in the prescribed dimensions of catheter tubing, and one company was stuck with several thousand feet of pediatric-size tubing spooled up like telephone cable. The vendor was prepared to take a loss on its overstock, but converting it into a tax-deductible donation was far more attractive.

Product changes and redesigns were equally common. The supply nurse at another hospital explained, "We had been using saline intravenous solution packaged in plastic bottles for years, but then the supplier decided to phase out the bottles in favor of plastic pouches. Since slightly different attachments were required for the pouches, the vendor was willing to give us everything we needed to get us to use the pouches instead.

"We agreed, and the switch was made. But there was still one problem. What about the entire pallet of IV solution in plastic bottles that we had in our warehouse? The vendor ended up crediting us for the backstock, but the company certainly didn't want the bottles back. If we couldn't find someone who wanted it, one hundred and twenty cases of IV solution in plastic bottles would be headed for a landfill site."

New, but no longer usable stock wasn't the only source. Surgical nurses at another Peninsula hospital turned recycling into an art form for the children of Chernobyl.

"Surely this only happens in America," one of the surgical nurses explained. "The hospital bean-counters determined that it was cheaper for surgeons to use disposable stainless steel surgical instruments, packed in surgical 'kits,' and throw them away after the kit was opened than it was to pay the cost of having reusable instruments autoclaved and stored.

"But we couldn't bring ourselves to be so wasteful, especially when we knew that stainless steel instruments were 'beyond price' in some parts of the world. So, on our own time and without any cost to the hospital, we started disinfecting the used instruments and packing them away with the items in the kit that had never been used at all."

When Michelle passed along a newspaper clipping about some Soviet hospitals that were doing surgery with resharpened razor blades because they had no scalpels, the surgical nurses nearly doubled their output for the project.

Surprisingly quickly, a small room behind the organ pipes in the Congregational Church of Belmont began to fill with donated supplies. Michelle began to have some sleepless nights as she thought about the other project that had collapsed because the organizers couldn't get the supplies to the hospitals in Romania.

"What if we have all these medical supplies stacked in the organ closet and we can't get them to Minsk without paying the usual freight charges? It would take all the money we've raised to pay for the shipping; nothing would be left to buy the methotrexate."

Dr. Olga had told Michelle that there was a doctor in Frankfurt, Germany, who would be able to receive shipments of medicine and medical supplies and could be trusted to get them to her. She had given her the doctor's name and telephone number. So Michelle tried that option first. She telephoned the doctor, who said that he did indeed have a pipeline to Dr. Olga that could be used—if she could get the supplies to Frankfurt.

About this time another angel surfaced. Mike Venturino, a member of the church and a good friend of Michelle's, also happened to be a United Airlines pilot.

"Why don't you let me write a letter to one of the United vice presidents I've worked with, describing what you're trying to do?" Mike offered. "He might be willing to commit the resources of the airline to the project."

Answered prayers came in the form of a February 20 letter from United VP James Guyette: "United Airlines would be grateful if you could arrange that we be the carrier chosen to assist in the movement of medical supplies [to the Children of Chernobyl]. It is unfortunate that children have to suffer due to the Chernobyl catastrophe and we want to help assure they receive proper medical treatment."

He authorized shipment of one thousand pounds from San Francisco to Frankfurt, and Michelle had the shipping connection she needed.

That same week she received an invitation to attend a luncheon that industrialist David Packard, the cofounder of computer giant Hewlett-Packard, was hosting at his Los Altos Hills home. The luncheon was intended to introduce selected local editors to the Lucile Salter Packard Children's Hospital at Stanford University, a state-of-the-art institution serving a wide range of children, from those who needed tonsillectomies to those with life-threatening illnesses such as leukemia and other forms of cancer. The hospital, which was about to open, was built after the establishment of an initial endowment of fifty-two million dollars from the Packard family. The gift was in memory of Packard's wife, who helped plan the hospital before her death.

Michelle took the opportunity of the luncheon to talk about Dr. Olga's hospital in Minsk and the Children of Chernobyl Project. Then, when she got back to her office, she wrote a note to David Packard thanking him for the luncheon. She added, "I am going to be so bold as to include a flyer and an article about the children of Chernobyl, an issue close to my heart after a 12-year relationship with the people of the Soviet Union."

A short time later, she received a letter from Packard telling her that he had forwarded her letter to the Packard Foundation with the recommendation that a small grant be made. She was thrilled—and a little naive, since she hadn't a clue as to what was involved in applying for a grant.

The foundation director contacted her next, and said he was suggesting a small grant of about five thousand dollars (which would have more than doubled the amount already raised). "But we need a variety of documents to establish the church's non-profit status; the tax-exemption number that the church puts on all its tax documents isn't enough."

That was the beginning of a month-long paper chase to find the appropriate original letters from the Internal Revenue Service and the office of the Northern California Conference of the United Church of Christ.

Meanwhile, another member of the November peace delegation, Gloria Bordeaux-Pacholec, had rallied the Sunday School of the First Congregational Church of San Jose to the children of Chernobyl cause.

"We picked Dr. Olga's hospital to receive the proceeds of the Sunday School's annual Mission Outreach program," Gloria wrote. "The children sold home-baked cookies, cakes and candies several Sundays after church and collected pennies for the project.

"One Boy Scout who had been saving money for a trip to Alaska donated his savings. Within six weeks, the children raised $1,038."

The check added to the growing fund at the Congregational Church of Belmont.

❊ ❊ ❊

ON FEBRUARY 5, a fax message reached the Moores in New York. It was from Vladimir Lipsky of the Byelorussian Children's Fund: "With the feeling of great sorrow and regret, we must inform you about a sad piece of news. Today Natasha Ptushko has died.

"Her state was getting worse and worse these last days of hers. The thin string of her life was broken at 17:00" (5:00 P.M.).

PJ was on his way home from college. When he arrived at the house and Paul broke the news to him, he sank to the floor. His father held him while they both cried.

Later Michael and Rebecca received a letter from Nadia Vyalkova, the interpreter they had come to know and love on the trip. "You know about little Natasha's death. Two weeks have passed since that time, but I cannot regain my spirit and cheerfulness. . . . I see her only alive. I ask myself and do not find an answer why it was she who was chosen by God to leave our earth for heaven. . . .

"Unfortunately, death is our routine. It is an inseparable part of our everyday life, but this case is very special and very hard, very hard indeed."

PART THREE

Little Vova, just six years old, could hardly contain his excitement when he found a snake sunning itself on a rock by the lake at the edge of the camp. Although he was chronically short of energy because of his illness, he ran panting up the hill to drag Gib, the nature counselor, back to the rock.

Beaming with pride at his discovery, he watched, entranced, as Gib caught the snake and held it up so Vova could stroke it — at an appropriate distance, of course. Vova dug a shoe box out of the trash and held it open for Gib to put the snake in. Since Gib didn't speak Vova's language, it took all of his nonverbal communication skills to convince Vova that the snake needed and deserved to be free. They put the snake on the sand and crouched down to watch as it disappeared under a railroad tie that was serving as a lakeside bench.

The swimming pool was another unique attraction. None of the Chernobyl campers could swim, but bright orange life preservers kept them afloat in the deep end. Eleven-year-old Misha, who had been sullen and detached, cracked his first smile of the week as Brian fitted a snorkeling mask over his head. After a pantomime demonstration of proper technique, Misha flopped face down in the water and kicked for all he was worth. They had to drag him out of the pool when it was time for dinner.

Some of the American campers were surprised when the visitors turned down American kid classics like hamburgers and fries in favor of fresh bananas, strawberries, and melons. Dark-eyed Katya offered an explanation for turning her nose up at pizza: "It doesn't taste very good and it is a silly thing to do with perfectly good bread and cheese."

9

"I'VE ENDURED IT
ALL FOR YOU"

"WE GOT THE FAX early in the morning last Wednesday," Paul Moore's deep baritone voice began on his weekly radio broadcast in New York. "The news that Natasha had died hit the Moore family like a ton of black coal, as if a dump truck had backed up to our lives, covering us with dust and ashes. . . .

"Just days before, we had managed to get an invitation from New York University-Cornell Medical Center for Natasha to come for a bone-marrow transplant. Dr. Jim Bussel had agreed to do her diagnostic workup and arranged for a bone-marrow specialist to prepare her for the transplantation surgery.

"By then we had learned that when we met her, Natasha was already in the fourth stage of non-Hodgkins lymphoma," Paul said. "Perhaps even then it was too late.

"Between our first trip—when we met her—and Sharon's trip, Natasha developed acute leukemia, and cancer was found in her lungs. But then she began to show some signs of recovery when we came at Christmastime. We had new hope because she rallied and made it through Christmas. She gained some weight and started to walk again after four months of being unable to stand up.

"We were hopeful that this was the window, the remission, we had all prayed for. But almost immediately after we returned home in January, Natasha caught pneumonia and her lungs filled with fluid. Her weakened immune system was unable to fight the infection any longer. . . .

"What were we to make of her death in light of our almost constant prayers for her recovery?" Paul asked. " 'God's will' seemed like an incomplete answer."

Sharon was more philosophical as she joined the broadcast. "I know that there is another world after this life. And, though I would have written the script differently, I have placed Natasha in God's hands. I know she has eternal life and died knowing that she was loved.

"I cannot turn off my love for all the others because we have lost one. As we grieve for Natasha, we must find hope and medicine for the others, for Andrei and Antonina and Galina."

But Paul Jr. was reeling. He had been wrestling with the hard questions of "Why?" His parents tried to get him to talk through his grief, but most of the time it seemed to engulf him.

"I prayed for her recovery more fervently and more eagerly than I've ever prayed for anything," he told them. "After her story was told on the air last fall, tens of thousands of people were praying for her to get well. And still God let her die.

"I feel powerless and weak and I'm beginning to doubt," he continued almost inaudibly through his tears. "The only thing I ever wanted to be was a minister, but maybe it was me. Maybe I was the reason she died. Maybe I hadn't believed enough. I even think now and then that I shouldn't be alive now that she isn't."

PJ was particularly upset that the Russian Orthodox Church had refused to give Natasha a Christian burial since she had never been baptized in the church.

"But she believed in Jesus and she took communion. I know because I served it to her. When we were in Minsk in January, I saw her every day. Nadia Vyalkova came along and interpreted for us. We listened to my tapes of Christian music and sometimes I read from the *Book of Common Prayer.*

"I told her all the old Bible stories that kids learn and one day I told her the story of the Last Supper. I said the words 'in remembrance of me' and explained what communion means to a Christian. Finally she asked me to give her communion. The next day I brought the already-blessed elements with me and she ate the bread and drank the wine 'in remembrance of me.' "

The day PJ left Minsk to return to New York with his parents, he had gone back to the hospital one last time. "I told Natasha to have faith and smile at the future. I told her to trust in God and she would be healed. She said she did believe and then she asked me to take care of her little brother Dmitri who's sick too. I kissed her good-bye. She was so beautiful, but that was the only time I ever kissed her."

Still, Natasha had died. PJ hadn't been able to save her. He wanted to go back to Minsk so he could tell her he was sorry, so he could tell her mother, her sister, and her brother.

AN INVITATION FOR PJ to return to Minsk for an international youth conference in March arrived just a few weeks after Natasha's death. He knew this was his chance.

Once he got there, he went through with most of the meetings on the agenda of the four-day conference. Invited to speak at one of the sessions, he related a conversation he had had with a television interviewer when he arrived in Moscow just a few days before. "He asked me if God was punishing Byelorussia with Chernobyl. The person I used to be would have said, 'You bet!' But now I said that God is there for the Byelorussian people and he has taught me to be present in their pain as well."

The real reason PJ had come was to find some peace, some closure, in the death of Natasha. He went out to the village where the Ptushko family lived, knelt by her grave, and said goodbye. He described that journey to Citihope radio listeners when he returned.

"We got on this big, Greyhound sort of bus and drove out on these country roads for about an hour and a half to get to the village. The driver didn't know the roads. It was only late afternoon but it was already dark and we got lost. At one point we stopped and asked a villager for directions to the Ptushko home. He said, 'Who are you?' I told him, 'Paul Moore, Jr.' Then he said, 'Oh, you've finally come.' And he told us how to get to the house. . . .

"Natasha's mother Alla was crying and she kept stroking my hair. She said I had hair just like Natasha's. Then Natasha's twin sister came in. I had met her once before in the hospital and Natasha had told me, 'This is what I looked like before I got sick.' It was weird seeing her twin standing there, almost as if Natasha was alive again.

"We went out to the gravesite and it was dark and raining. It was just outside the village and there was no gravestone, just a latticework frame filled with flowers covering the whole grave.

"I remembered that a Swiss doctor who was working at the oncology hospital and who had been staying at our hotel wrote me that Natasha had stood up outside her bed before she died, that she had said she wasn't afraid and that she'd wait for Paul—for me.

"We had a proper Christian funeral for her right there in the rain. I said that the disease hadn't beaten Natasha. She had stood up at the end. Her soul was strong—she'd been telling those Bible stories to other children and reading her Russian Bible to them—even when her body couldn't go on.

"I said I would never forget her or her powerful, pale blue eyes. They were burned into my memory.

"As we walked slowly through the rain back to the big bus, the interpeter, Nadia Vyalkova, broke down, and I was left to say my goodbyes without her. But I didn't need an interpreter to understand what Alla wanted as she hugged her three-year-old Dmitri and pleaded with me to help him now that it was too late for Natasha.

"And I will," PJ promised. "I'm absolutely committed to that. I'll bring him to New York by myself if that's what it takes."

❊ ❊ ❊

WHILE HIS SON WAS IN MINSK, Paul Moore was working out the details of bringing Zorachka, the Byelorussian children's folk song and dance group, to the United States. He was planning an eight-city Passover/Easter tour that Citihope was prepared to sponsor as part of the fifth anniversary observance of the disaster at Chernobyl.

Michael Christensen had been busy since the group returned in January, lining up local sponsors and hosts for the thirty-seven-member group that would spend a week in New York, New Jersey, and Pennsylvania.

"I see a couple of ways for Citihope to raise money for the Chernobyl project with the Zorachka tour," Michael reported to Paul by phone from his home in San Francisco.

"Besides the proceeds from tickets, we can sell Byelorussian crafts, like the straw ornaments and lacquer eggs, at the site of each concert. The money from those sales would go directly to the programs of the Byelorussian Children's Fund.

"Any additional pledges we get during the concert would go to help fund the mobile diagnostic clinic."

The two-hundred-thousand-dollar mobile clinic was still a high priority for Citihope. Once it was purchased and outfitted, it would be used for monitoring children at risk and for some forms of treatment in the contaminated Gomelskaya region of Byelorussia.

"I don't have any problem with that," Paul said, "but I really think we need corporate sponsorship. I think I might be able to get Pepsi-Cola to go for it.

"And we probably should expand the invitation to include five chaperones, and two of those youngsters who did those fantastic drawings of (the contaminated city of) Narovlia." Once again Paul was thinking on a grand scale. "Then if we include five representatives of BCF, we'll have a total of forty-nine. I don't see any reason why the original dates of March 29 through April 8 won't work. There's time to pull it all off."

Citihope had assembled an elaborate publicity kit including posters, camera-ready bulletin inserts for churches, color photos, and a videotape of Zorachka.

"I think we've finally lined up enough performances to make the trip worth the BCF's hard-currency investment in airline tickets to get them all to the United States," Michael said. "I know that's a real financial burden for them."

And that's where the whole ambitious project foundered. The airline tickets never materialized; Zorachka never left Minsk; and Citihope was left holding a week's worth of soon-to-be-unfulfilled promises.

Paul was scrambling. "There's no way I'm going to let this all fall through. If it does, we won't ever be able to go back to those sponsors for help again. We may have to be a lot less ambitious, but there is going to be a tour."

Free-lancing by fax and telephone, Paul came up with airline tickets and reservations for four. "I'm going to call Elena Dmitrieva (director of the Pioneer Palace that sponsored Zorachka) and ask her to choose two from the thirty-seven kids in that troupe to come to New York with her and Nadia Vyalkova. I can't get hold of either Lipsky or Trukhan (of the Byelorussian Children's Fund) to approve the trip—they're out of the country or something—so I'll issue the invitation from Citihope. This way we can have these four make all the promised appearances with just a slight adjustment of dates."

Sharon Moore would be faced with a serious rift to mend with BCF when she arrived in Minsk later that month, but the first week in April, the Zorachka tour went on as planned, albeit on a considerably reduced scale.

The two members of Zorachka who made the trip, Nadia Ivanova and Misha Boyko, proved equal to the task of representing the entire troupe. Dressed in their traditional Byelorussian costumes, white linen embroidered in brilliant red in the familiar geometric patterns of their Slavic ancestors, the teenagers provided the core of the performance on which Citihope was able to hang a successful appeal. The entire Zorachka ensemble did perform—via cassette tape —one song that they had practiced in English.

ONE OF THE MOST receptive stops on the tour was the Hutterian Community in Pleasant View, New York. Here, three hundred and fifty brothers and sisters live in a commune that supports itself with its own educational toy factory. At a press conference after the performance, Michael took out his handheld geiger counter and held it up in the communal dining room.

"This interior registers 12; it's always a little higher than that in New York City," he explained. "But at many places we visit in Byelorussia, this meter reads 380 or higher—which isn't a problem if you only stay a little while. But if you live in that environment day in and day out all your life, there can be serious, life-threatening effects—especially for the children."

Blond and blue-eyed Misha proved to be a particularly eloquent spokesperson for the children of Chernobyl. In excellent English, the fifteen-year-old described the uncomfortable reality of growing up in Byelorussia. "Of course, I hope I don't become sick, but it's impossible to pretend there's no threat. I took part in a big May Day demonstration outdoors just days after the catastrophe."

Elena Dmitrieva added an emphatic condemnation of the official response to Chernobyl. "Our government's worst mistake was choosing not to warn us to stay indoors with windows and doors sealed shut during those first days. If they had warned us, the exposure would have been less. There wouldn't have been innocent children walking in the streets, many barefoot, on this very hot May Day after the explosion."

While visiting the Hutterians, Michael somewhat naively suggested to Elena that this self-contained pacifist farm community, which made decisions by consensus and shared all its resources, was an American version of the Soviet collective farm he had visited, Vu Pokalyubichi.

Dmitrieva replied with a bitter laugh: "Oh, Michael, don't you see? These people want to be here. They chose this life and they can choose to leave. It's very different from those on the collective. Their parents and grandparents were driven from their own farms and their freedom to choose was lost."

With the Hutterians and at each of the other stops on the tour, Michael took every opportunity to announce that in November, the entire Zorachka ensemble would make the tour they weren't able to make in April. Citihope was gathering sponsors for the expanded nineteen-day fall tour, but this time they were covering their bases. Paul and Michael had given the BCF a June deadline for reserving airline tickets so they would have ample time to promote the tour—with confidence that Zorachka would be able to come.

When that deadline passed with no airline tickets in hand and the hard currency situation in the Soviet Union plummeting, the November preparations were put on hold. But Paul wouldn't give up on Zorachka. That group of talented Byelorussian youngsters was the perfect vehicle for carrying the story of the children of Chernobyl to America. Now he would look for a corporate sponsor who could pick up the entire transportation tab.

❊ ❊ ❊

IF THE CHILDREN'S ensemble couldn't get to America in April, Citihope would go to them. As part of the Byelorussian republic's observance of Chernobyl Anniversary Week, Sharon Moore had been invited to take part in an international charity concert on April 24 in Minsk.

She left New York with Citihope's sixth shipment of oncological drugs, antibiotics, vitamins, and food for the BCF. A former professional singer, she would be on the same concert program as Soviet and Western rock groups, including Pink Floyd.

"I'm willing to sing," Sharon told the promoters in Minsk, "but I want Zorachka or Peace Child to perform my first number with me."

She got her way.

A sound stage, complete with every possible piece of amplification equipment, was constructed at one end of the huge sports stadium in Minsk. The concert was set to begin at dusk on a very cool, crisp day. Because of the cold, the concert organizers decided to bring the children—and Sharon—on first.

In an internationally broadcast production before twenty thousand Byelorussians, Sharon brought her message of hope and America's compassionate response to the children of Chernobyl, surrounded by thirty of the most talented of those children. The response was overwhelming—so positive that the producers invited her and the children back to open the reprise of the concert the next night.

At the conclusion of Sharon's performance the first night, a very special young man appeared at her side. Carrying an armful of flowers for her was Andrei Aleksandrovich, the seven-year-old whose mother had appealed to Sharon for help at the Palace of Pioneers in Minsk in the fall, the one whose bone-marrow slides were too old to be of any use in diagnosing his condition.

Andrei had more than flowers to give to Sharon. He put into her hand fresh bone-marrow slides, carefully wrapped in blue tissue paper and rubber bands. He told her, "I've endured it all for you, madame."

He bowed most dramatically, and Sharon knelt down and enveloped him in a bear hug. "I promise to carry these hard-earned slides to the doctors at Cornell as soon as I am back in New York. You have been a very brave soldier, Andrei. So very brave."

✳ ✳ ✳

THE CHERNOBYL ANNIVERSARY observance occurred during a brief lull in a season of public unrest in Byelorussia. Widespread discontent over Chernobyl and a slew of economic issues offered the Popular Front a new opportunity to flex the political muscle it had first discovered in the Revolution Day demonstrations the previous November. Workers' strikes in Minsk and eight other Byelorussian cities put the Communist government of Nickolai Dementei on notice that the republic would be quiet and acquiescing no more.

"The initial work stoppage erupted spontaneously in Minsk on April 3," Nadia Vyalkova explained to Sharon during her visit. "It was a protest after Moscow followed through with its plan to institute dramatic price increases on most consumer goods and services. But the strike quickly became more political. People showed up carrying hand-lettered banners reading 'Gorbachev and Dementei step down,' 'The Communist party drove us to this,' and 'Put food from Chernobyl on the government table.'

"By the next day, students, doctors, and government workers had joined the strike. The white-red-white nationalist flags, which were banned six months ago, were everywhere.

"I think the government was surprised at the depth of the people's anger—and maybe a little scared. But, for whatever reasons, they granted television airtime for the strike committee to broadcast its demands."

That list of demands included higher wages to keep even with the newly inflated prices, repeal of the five percent national sales tax (called the President's Tax), the removal of Communist party cells from factories and the party from government institutions, the resignation of Gorbachev and Dementei and their administrations, multiparty elections for the republic's parliament by July, a fifteen percent reduction in income tax, and an end to special privileges for the party *apparatchik* and *nomenklatura*.

This was heady stuff for a republic with only a modest history of political unrest, but the strikers were just hitting their stride.

"On April 9," Nadia continued, "Gorbachev called for a moratorium on strikes; the next day two hundred thousand people turned out in defiance in Minsk."

The strikers also had some new, more ambitious demands: nationalization of Communist party property, a sharp cut in factory contributions to the central Soviet budget, round-table talks with all the political parties in the republic, transfer of power to a coalition government, and the legalization of private ownership of land.

Despite the depth and strength of the uprising, Dementei had no intention of folding his tent. With federal troops massed on the side streets of Minsk, he told the strike committee, "When a patient is sick and he is put on a surgical table . . . his arms must be strapped first so he won't interfere when he starts feeling pain. Reform of our society, changing from one state to another, namely to a market system, is painful surgery."

"He talked tough," Nadia said, "but the troops were never ordered out. Then on April 11, the government agreed to open talks on the strikers' demands. The strike was suspended—but not for long. When Dementei went back on his promise of negotiations on April 23, workers rushed into the streets with even stronger anger and frustration."

Tension was thick, and bloodshed was threatened when workers tried to block the main rail link from Moscow. Then, on the anniversary of Chernobyl itself, the Byelorussian parliament acquiesced and set up a committee to meet with the strikers on the list of demands—one of which had already been met: Gorbachev had canceled the detested President's Tax.

A brief revival of the protest occurred in May, when the parliament refused to consider the strike issues, but by then the committee had adopted new "go slow" tactics and the strike was effectively over.

But the Minsk rebellion had achieved its most significant goal. Gorbachev was put on notice that he could not count on even the most conservative and historically complacent of the Union's fifteen republics as he struggled to hold the disparate elements of the Union together.

10

TAPPING THE PIPELINE

IN THE FIRST DAYS of that anniversary month, Michelle Carter's Children of Chernobyl Project overflowed the storage room behind the organ pipes at the Congregational Church of Belmont. She was, again, lying awake nights worrying about getting the medical supplies to Dr. Olga in Minsk.

"I called that doctor in Frankfurt again, the one who had agreed to get them from there to Minsk," she told the two people who were working most closely with her on the project—her husband Laurie and friend Mike Venturino. "But he's on sabbatical until September.

"I certainly don't want to wait until then. United will fly them to Frankfurt, but United doesn't fly to Moscow and, for sure, not to Minsk. And there is no guarantee they will arrive at all unless the supplies are escorted. I have no intention of going to all this work and then having everything diverted to the black market, or sitting in an unheated warehouse until their usefulness is wasted."

This time it was Joyce Weir, a dental hygienist from the state of Washington and a member of November's United Church of Christ delegation, who provided the answer. Joyce had read an article by Rebecca Laird-Christensen about an organization called Citihope that was involved in a compassionate response to the children of Chernobyl. Joyce suggested that perhaps they could help in getting the Belmont church's supplies to Dr. Olga.

Joyce had tracked Rebecca down at her home in San Francisco and learned that Citihope had a representative there, Rebecca's husband, the Rev. Michael Christensen.

Grasping at a possible solution to her shipping crisis, Michelle called Michael, and two parallel paths finally converged. It was the first time that either knew about the efforts of the other, and there was no instant chemistry when they met over lunch at a small Thai restaurant near Michael's home. They talked first about the issue at hand—transporting Michelle's medical supplies to Dr. Olga.

"Citihope is taking a huge shipment of antibiotics, vitamins, methotrexate, and supplies to Minsk in April as part of the observance of the fifth anniversary of Chernobyl," Michael said. "Sharon Moore will deliver the shipment personally. There's no reason why your medical supplies can't be included with the Citihope cargo—if you can get them to New York in time."

They talked business, but each was also measuring up the other. Michelle had never heard of Citihope, and her reporter's natural suspicion kicked in. She asked a lot of questions about how they raised money—and spent it. Her antennae picked up on Paul Moore's "radio ministry" when Michael mentioned it, and she stifled a moan.

Dear Lord, tell me I'm not getting involved with some televangelist who's only interested in self-promotion. I may send some of our supplies, but they aren't getting a penny of our money until I know a whole lot more.

Yet Michelle was impressed to find out that Michael directed the United Methodist AIDS Project, served as a chaplain on the AIDS ward at San Francisco General Hospital, and had also founded Oak Street House for the homeless. She had heard about the good work done there. However, she was having some trouble reconciling this activism with his conservative, Nazarene background. She had never thought of Nazarenes as being in the forefront of social action.

Just when she was beginning to pull back, Michael said exactly the right words: "Maybe you ought to come along on one of the trips to see how we work."

That's what Michelle wanted to hear. She didn't trust people with too many secrets, and she was dealing with money that a lot of caring people had entrusted to her. She didn't want to make any mistakes.

As cautious as she was, she couldn't help but like Michael. He was open and interested in her experiences in the Soviet Union. They laughed about the spontaneity and flexibility required in dealing with the Soviets and the inability to plan almost anything in advance.

When Michelle left, she felt she'd made a friend and perhaps forged a partnership for her Children of Chernobyl Project. For Michael, the telephone call from Michelle was just another of many inquiries from people who had heard of Citihope and wanted something from him. But Michelle had mentioned her church's Children of Chernobyl Fund; she had raised money to buy medicine. Potential donors and benefactors always got Michael's attention.

Michael took note of how serious Michelle was about her stewardship over the money she had raised. Not a penny of her fund went for overhead or staff salaries. All of it was to be spent for medicine—and not just any medicine, only methotrexate for Dr. Olga Aleinikova at the Children's Hematological Center in Minsk. Michael had never before seen such specificity in fundraising, and zero percent overhead and administrative fees was unheard of.

As they exchanged ideas that day, neither Michael nor Michelle had the slightest notion of how closely they would work together in the days ahead.

❊ ❊ ❊

MICHELLE'S MIND was racing after she left the restaurant and drove south along the west side of San Francisco Bay to San Mateo. She had less than a week to get all the medical supplies inventoried, packed, weighed, and delivered to San Francisco International Airport.

She made a mental note: "I'm sure that United Airlines won't balk at shipping the cargo to Kennedy International in New York instead of Frankfurt, but I'll ask Mike Venturino to set everything up."

A call to Citihope set a date of April 19 for her to have everything in New York so it could leave with the April shipment, and she wanted to give United time to move the cargo at their convenience. Everything would have to be packed by Sunday night and ready for her to deliver to the airport during her lunch hour on Monday.

After church on Sunday, Michelle, Laurie, and Mike gathered packing materials, tape, a bathroom scale, and a spiral notebook in the fireplace room in the basement of the church and started unloading the organ closet.

"I don't think it's smart to send everything we have on this first trip," Michelle said. "Let's test the pipeline and see how well it works. If everything goes smoothly, we'll be ready to send a really big shipment with the next Citihope trip in June."

They sorted out thirty-five hundred sterile syringes and needles, thirty dozen IV sets and IV extensions, several hundred pediatric catheters, boxes of surgical tape, disposable stethoscopes, sterile drapes, and iodine applicators. Also included was a pipette washer with racks and stain jars for the lab at the Hematological Center, donated by the Redwood Medical Clinic in Redwood City, and an eye-ear-throat examining instrument from a doctor at San Mateo County General Hospital.

It was a modest shipment of eight cartons—each with an itemized inventory taped to the outside and covered in clear plastic. "Children of Chernobyl" was written in large black letters on the side of every box, just as Dr. Olga had instructed. Weighing a total of two hundred pounds, the cargo wouldn't make much of a dent in the thousand pounds that United had promised to ship, but it would be a good test of the system.

United was as good as its word. On April 16, Citihope was notified that the supplies were at Kennedy, waiting for transshipment. Those eight cartons from the Congregational Church of Belmont were in Minsk before the fifth anniversary observance of the catastrophe. That was just five months after the peace delegates from the Northern California Conference of the United Church of Christ had listened to Dr. Olga's impassioned plea.

In June, ten volunteers would be needed to pack the forty-one cartons weighing more than six hundred pounds. Those boxes included ten thousand sterile syringes, needles, catheters, blood collection tubes, and nearly a thousand carefully packed stainless steel surgical instruments that nurses from a local hospital had rescued from their "use once and toss" destiny.

"You really ought to think about going on the next Citihope trip," Laurie told Michelle. "We know the pipeline works, but it would be worth seeing them in action."

Laurie was content to play a supporting role in the project. After nearly twenty-five years of marriage he was used to life with an activist wife. He lent support when it was needed and was happy to step aside and let Michelle carry the ball when it wasn't.

"Oh, I'm going all right," she replied. "I'm really curious about how this organization works."

❊ ❊ ❊

ONCE SHE HAD CONFIRMATION that the first shipment of medical supplies had arrived at Dr. Olga's hospital as part of Sharon Moore's Chernobyl anniversary trip, Michelle was ready to consider spending the seven thousand dollars or so that had accumulated in the account at the church. It was no secret what she would buy—methotrexate.

She picked up a copy of the Physician's Desk Reference on pharmaceuticals and started dialing the companies listed under the description of methotrexate. It took only a couple of phone calls to discover she was spinning her wheels. She couldn't get past the receptionist at these huge pharmaceutical firms. They could recognize a "touch" coming a thousand miles away, and she was getting nowhere.

Once again she sensed she was trying to reinvent the wheel. She needed to tap into an established pipeline, and where better to turn than Michael Christensen? He had the name and number she needed at his fingertips—Paul Maxey of Interchurch Medical Assistance, Inc. He was the one who had come up with the eighteen hundred vials of methotrexate the previous fall so Paul Moore could fulfill Dr. Olga's requirement for a miracle. Surely he could handle a request for the three hundred and fifty vials that her fund's seven thousand dollars would buy.

As it turned out, Paul Maxey held the key to far more than methotrexate, and his contribution was vital to her church's small Children of Chernobyl Project, as well as to Citihope's vast effort. He put Michelle's seven thousand dollars to work. When she arrived in New York in June to take part in Citihope's delegation to Minsk, three cartons of methotrexate were stacked in Citihope's crowded Manhattan offices. They were ready to be loaded for the journey to Minsk, so that eventually Michelle could carry them in her own arms through the dim, paint-peeling halls of the Children's Hematological Center and stack them next to Dr. Olga's desk.

She would be glowing, partly from the physical exertion of hauling cartons in the June heat—but just partly.

11

"WE CAN'T
BE RESCUERS"

THAT APRIL Svetlana Khalko, a ten-year-old leukemia victim, was brought to the United States through the efforts of Sharon Moore and Citihope.

Originally, the Moores had been looking for a hospital willing to treat Natasha Ptushko. Dr. James Bussel at Cornell Medical Center-New York Hospital had agreed to waive his fees in treating her. Arrangements were proceeding when Citihope received word of Natasha's death.

The Moores chose to continue their efforts as a testament to Natasha, and they invited the Byelorussian Children's Fund to submit the name of another candidate. They sent medical records and bone-marrow slides for two children. Dr. Bussel examined both sets of documents to select the child most likely to benefit from treatment in the United States.

Svetlana was that child. She had first become ill in December 1990. Her initial symptoms—high fever, swollen lymph nodes, and lethargy—had been treated with hot compresses until doctors in Minsk pronounced the diagnosis of leukemia. She never responded to treatment in Minsk and was quite ill when Citihope brought her to New York.

Svetlana arrived on Good Friday. Paul and Sharon Moore brought her and her mother from the airport to their home in New York. On Easter Sunday they all attended services at St. George's Episcopal, where Paul introduced the plucky young girl to the congregation.

The arrangement with Cornell Medical Center-New York Hospital was that Svetlana would be treated as an outpatient. Citihope found housing for her and her mother in Manhattan and continued to support her while Dr. Bussel did his evaluation.

However, it turned out that Svetlana was much sicker than anyone had realized. She had to be hospitalized almost continuously from the time she arrived in the States, quickly running up an eighty-thousand-dollar hospital bill that Citihope is still struggling to pay.

Dr. Bussel was blunt: "She needs a bone-marrow transplant or she will die. Frankly speaking, the bottom line is money. If you can raise the funds, I can make the arrangements right away."

He explained that the transplant surgery itself costs about two hundred thousand dollars and requires a one hundred-day convalescence. Doctors generally give the procedure a fifty to sixty percent chance of success.

Sharon launched another telephone crusade, but met nothing but brick walls until someone gave her the name of Joan Keller of the National Bone-Marrow Registry in Washington, D.C.

"She was the miracle worker," Sharon said. "She called me from Hawaii where she was vacationing and said she would take on this case. She contacted the Fred Hutchinson Cancer Research Center in Seattle and the Olga Korbut Foundation and they agreed to take Svetlana as their first advocacy case."

Doctors at the Hutchinson Center evaluated the status of Svetlana's acute lymphocytic leukemia and decided that a further course of chemotherapy was necessary before a bone-marrow transplant could be attempted. However, while Svetlana was being stabilized, blood-matching tests on her family members determined that her nineteen-year-old brother was a perfect match. He would be the bone-marrow donor for his sister.

Doctors in Beyelorussia were fully aware of the need to provide bone-marrow treatment for leukemia victims. But it was very difficult to establish the totally sterile environment that the procedure requires, given the limits of their treatment facilities. For this reason, Dr. Olga Aleinikova had included a separate, sterile transplantation unit in her plans for a new children's cancer center in Minsk.

* * *

THE CASE OF SVETLANA was the first part of a broad learning curve for Sharon Moore and Citihope. "We took in Svetlana," Sharon observed, "as we had planned to take in Natasha—to rescue her, without much thought as to how everything would be paid for. We had verbal commitments from Dr. Bussel, but we didn't realize that all he could do was agree to waive his fees. He couldn't speak for Cornell. When it became clear that Svetlana was much sicker than we thought and would have to be hospitalized the whole time, Cornell gave us a reduced rate, but they still expected us to pay."

As Citihope moved through the process of handling other advocacy cases, Sharon learned that an emotional "rescue" response had to be replaced with "wise facilitation." Written commitments from doctors, hospitals, and support organizations had to be in place before a child could come.

"Most of all, we learned not to take on responsibility for the child," Sharon said. "We would do the work of obtaining commitments from the people who would be responsible. Never again would we make the mistakes we made with Svetlana. We are facilitators; we can't be rescuers."

Citihope had learned something about how to make the right contacts when PJ Moore invited fellow teenager Vladimir Luchevisti to come to the United States for treatment of bone cancer in his left leg. He faced amputation as the only treatment available in Minsk.

Vladimir's mother and father were committed to saving their son and his leg. On their own they had raised five thousand dollars from factories and enterprises in Byelorussia. They had also obtained the sponsorship of the Byelorussian Children's Fund, which made the initial contact with the Mayo Clinic in Rochester, Minnesota. All the family really needed Citihope to do was confirm arrangements in Minnesota and serve as facilitator before and during Vladimir's stay.

The teenager and his mother arrived in New York in March. The Moores shepherded them through a short stay in New York and then on to Minnesota, where the Ronald McDonald House—the charitable foundation established by Ray Kroc of the McDonald's fast-food chain—agreed to provide housing.

Vladimir received aggressive chemotherapy and radiation treatment at Mayo, but still came very close to losing his leg. However, the tumor was sufficiently reduced that he was able to return to Minsk in December with his cancer in remission. Further visits to Mayo would be required to monitor the progress of his recovery.

"We felt very positive about our experience with Vladimir," Sharon said. "His case was difficult but successful, and again we learned how we could be most useful."

By the end of the year, Sharon had also lined up medical care and day-to-day support for a three-year-old heart patient (and her mother) who would be treated at Cornell Medical Center in early January.

AT THE OTHER edge of the continent, Michelle Carter would be able to deliver the news to the father of Danil, an infant boy with Pierre Robin cleft palate syndrome, that the international organization of plastic and maxillofacial surgeons, Interplast, would be willing to do the corrective surgery at San Jose Medical Clinic in California, or at any one of a number of sites in Europe. This baby would be the first child of Chernobyl for Interplast, which had already been involved in reconstructing the faces of children disfigured in the civil wars in Central and Latin America.

The appeal for help for Danil had arrived in a letter carried by a chaperone accompanying children who were to attend a cancer camp in the Sierra Nevadas of Central California. "I don't know this child, but the father asked me to give the letter and the photographs to someone who might be able to help," she said as she passed the appeal along to Michael Christensen.

Advocacy cases were the most difficult to respond to, not just because of the logistics and expense of bringing a child to the U.S. for treatment. They also raised moral issues that few Americans were prepared to deal with. The kids brought to the States were certainly only a tiny fraction of the Soviet children needing treatment, and in addition there were the unmet medical needs of many American children.

These children succeeded in getting treatment because they had someone who could break through the system and get attention for them, an advocate who knew whom to contact at the relief organizations in Minsk, and who could plead with Americans in English. They had someone to tell their stories and, if necessary, beg for them.

What about the others who had no one to take up their cause? Would they die? Perhaps.

The Americans had to insulate themselves against the overwhelming size of the task they had taken on and, occasionally, the guilt they felt for not being able to help every needy child.

❊ ❊ ❊

PROFESSOR ROBERT MILLER was convinced that the fifth anniversary of the explosion at Chernobyl would provide the perfect opportunity to find out who in the country was responding to the needs of the children, get them all together in one place, and compare notes.

As director of volunteer services for actor Paul Newman's Hole-in-the-Wall Gang Camp in Ashford, Connecticut, Miller had gained valuable experience the previous August when the camp hosted seven children of Chernobyl.

The camp, which treats children with cancer and other life-threatening illnesses, had opened in June 1988 after three years of planning. Newman took the name of the camp from the story behind his movie "Butch Cassidy and the Sundance Kid." He told the *New York Times* that he was launching the project "after I lost a few friends before they should have gone."

Built of rough timbers and totally handicapped-accessible, the thirty-five buildings of Hole-in-the-Wall Gang Camp were designed to handle about a hundred campers at a time, ages seven to seventeen, who attend free of charge. "I knew what I wanted to avoid, which was a sense of institution," Newman told the *Times*. "I wanted it to look like a turn-of-the-century lumber camp in Oregon."

But Disneyland it wasn't. Every building was totally functional, including an unobtrusive infirmary staffed at all times with a pediatric oncologist and nurses. Among the groups providing support services were the American Cancer Society, the Leukemia Society of America, the Elizabeth Kübler-Ross Institute, COPE, Compassionate Friends, City of Hope, Friends of Karen, and the Ronald McDonald Children's Charities.

Prior to Citihope's first visit to Byelorussia, Hole-in-the-Wall Gang Camp had issued an invitation to one of the Chernobyl relief agencies to send seven campers. No group had more experience in providing rest and recreation opportunities for children of Chernobyl than Hole-in-the-Wall Gang Camp—and only one or two others had done as much. Because the camp had such a high profile, Miller seemed to be constantly on the telephone, making suggestions or giving advice to groups that wanted to offer a camping experience or home visit for these children.

Within six months after the seven Ukrainian children left the Connecticut camp, an informal network of American relief organizations was emerging. It was time to pull representatives from all these groups together for a day-long seminar, and Washington, D.C., was the logical place.

In writing his proposal for the conference, Miller, who is also a professor of music at the University of Connecticut, said:

"Our experience tells us that these children may not be well-screened in the Soviet Union. Some may be medically unstable; some may not bring with them medicines to continue therapy while in the United States. Though well-meaning, some of the host groups . . . seem to be unaware of the special care and facilities that the children may need. They are unaware, for instance, that some children are at risk of becoming very ill very quickly.

"As well, they are unaware of the moral dilemma that they may face as they contemplate the return of children in mortal peril to a strained medical system in the Soviet Union, where substandard care and its terrible consequences can be predicted."

Miller had some specific goals he wanted the conference to achieve. "Let's see if it's possible to establish a medical screening or 'triage' procedure in the Soviet Union to make sure that campers are suitable candidates before they leave for the United States.

"I'd also like to try to develop a standard medical status form to accompany each camper who comes to the U.S. Another important goal would be the sharing of pharmacological data among medical professionals at the various camps. We know that American equivalents for Soviet medications often cannot easily be determined, and many local or regional medications being used by these children have no American counterparts.

"We also need to set up a procedure that will allow sponsoring organizations and ordinary camps to use the already-established Children's Oncology Camps of America network to accommodate sick campers if they need to.

"Finally I want to create and distribute a directory of services and interested parties. We need to know who we are and what we are able to do."

As the conference began to take shape, Miller discussed the idea with Paul Newman and an anonymous benefactor. They liked the possibilities. Within two weeks, the conference was scheduled and filled with key participants—strictly by word of mouth.

Michael Christensen was on the phone with a staff member at AmeriCares, trying to pry loose some funding for a mobile diagnostic clinic for Byelorussia, when he heard about the conference. AmeriCares gave him the telephone number of Hole-in-the-Wall Gang Camp, and Citihope was invited to participate in the conference.

Senator Thomas Dodd of Connecticut hosted the session in the Capitol on April 11. Two dozen "power players" from around the country showed up, and Michael was feeling just a little awed by the gathering. A preliminary estimate indicated that it would cost more than twenty-five thousand dollars to convene this one-day workshop.

"That's a lot of Newman's salad dressing," Michael figured, knowing that the proceeds from the sales of the product provided the principal support for the actor's charitable foundation.

The format of the conference was informal, with lots of questions from the floor. Connections were made on the spot. A speaker would say, "We need this," and someone in the audience would stand up and say, "We've got it."

One of the few other agencies with direct experience bringing these children to the United States was the YMCA of Michigan. It had brought ten Byelorussian children with cancer to a "Y" camp in Michigan the previous summer, and planned to do it again the following year.

Jerry Courtney of the Michigan "Y" and Victoria Rakowsky of the American Cancer Society of Michigan talked about what had worked for them and what hadn't. They advised host organizations to be prepared for language and cultural barriers, logistical problems, media intrusions, and medical complications, underscoring the need for absolute flexibility.

"Remember that the children will arrive exhausted, and a day or so of nothing but rest and getting acquainted should be scheduled before they can be ready for camp," Jerry said. "And the children—even the healthy ones, but most certainly the sick ones—should have a medical evaluation as early as possible."

They offered copies of their *Volunteer Handbook* and their *Medical Procedures Manual* to any group that could use them, and then Courtney ended with a warning: "Since medical treatment is very expensive in the United States, you need to face the fact that most of the children who come will have to go back to the Soviet Union to get, as best they can, the treatment they need. That's not an easy realization to come to, under any circumstances."

The issue of treatment in the West would be revisited a number of times during the day. Dr. Ed Baum of Chicago, the international medical advisor for the Ronald McDonald Houses associated with children's hospitals throughout the United States, brought profiles of a number of children with conditions that could not get appropriate care at that time in the Soviet Union. "You need to contact local hospitals and find out the possibilities of getting treatment for these children," Baum urged. "Ronald McDonald House will provide housing for the families if medical assistance can be arranged."

Nadia Matiwsky, executive director of the New Jersey-based Children of Chernobyl Relief Fund, promised, "Our people will do everything we can to help support the children brought to the United States for treatment—and their families. Also, with the help of numerous private and corporate donations, our group is working to build and equip a hospital in Lvov, Ukraine, to treat the victims of Chernobyl."

❈ ❈ ❈

THE NEXT SPEAKER was Dr. Molly Schwenn, a pediatric oncologist, who had been among the staff doctors when the August session of Hole-in-the-Wall Gang Camp opened. The idea of a camp for kids with cancer had appealed to Molly, a tall, no-nonsense outdoor type with a medical degree from Stanford and a teaching appointment at Harvard. She had known she wasn't in for a holiday when she arrived at camp, but she had also been expecting a break from academic medicine and her practice at Boston Floating Hospital.

When the one hundred campers from across the United States arrived, she had been on hand with the nurses to check medical information forms, dosages, and supplies so they could send nervous, protective parents on their way with some degree of comfort. The American children had had a wide range of cancer diagnoses, serious blood conditions that included sickle-cell anemia and hemophilia, and immunologic disorders. Molly's task had been pretty much as she expected—largely one of implementing the medical maintenance programs already mapped out for these children.

But late that same night, the seven children of Chernobyl arrived and put a serious dent in her expectations.

"The three girls and four boys, ages six to fifteen, had been traveling for twenty-four hours and they were exhausted. The doctor traveling with them, Theo Pedeshezky, gave me some scant medical information about each of them and a partial supply of drugs I'd never heard of.

"Neither the children nor Theo could speak a word of English, and my command of Russian was limited to a few words I learned as a third-grader in a Saturday morning class at the University of Wisconsin."

So medical histories had to be gathered with the help of Nadia Matkiwsky as translator. "I spent most of the next day examining the youngsters while nurses drew blood for blood counts and liver function tests," Molly said. "I was terribly impressed with the cooperation of the children. They were extremely brave, especially considering their language and culture shock. For the most part, they were generally OK—with the exception of dental caries (cavities), which were nearly universal."

Laboratory results arrived the next day and Molly reevaluated her new charges: Vladimir (also known as Vova) had a low white blood count, which made him additionally vulnerable to infection; Mikhail was anemic; and Katya had an abnormal liver test. All three had been receiving treatment for acute lymphoblastic leukemia (ALL).

"Then the real detective work began," Molly said. "Volunteer translators, organized by Bob (Miller), helped me drill Theo—who we discovered knew next to nothing about pediatric oncology—and translate the incomplete documents the children carried with them.

"We were finally able to piece together a sketchy history of treatment for each of the seven children," Molly recalled. "Five of them had leukemia, ALL, the most common childhood cancer in developed countries, and two had Hodgkin's disease, a cancer of the lymph nodes."

But one piece of the puzzle had defied a solution. "I had never heard of most of the drugs mentioned in the medical documents and hadn't the first clue where we could get them.

"I contacted a pharmacist with Soviet connections who called a number of places, including the Soviet embassy in New York and a variety of health food stores. We couldn't find any American equivalents, but we finally concluded that these drugs were homeopathic preparations and probably weren't essential for the kids' cancer care."

The children hadn't been concerned about the search for medication; they were far more interested in the fresh fruit, especially bananas, that were abundant at the camp and so scarce in the Ukraine. Molly had added a full complement of multivitamins to the chemotherapy that she prescribed for them and, over the ten days of camp, had kept a close eye on their infirmary visits that, more often than not, were prompted by bee stings and scraped knees rather than their life-threatening illnesses.

The Ukrainian seven had learned to use pantomime and hand signals to breach the language barrier, and the nature talks had been favorite breaks in the day.

As it came time for the children to leave, Molly could not shake her concern about two of them and the inadequate cancer treatment they had received before they arrived. In one case, she prepared an extensive written recommendation and gave it to Theo to take back to Byelorussia.

As for the other child, six-year-old Vova Malofienko, Molly was convinced that he would not get the appropriate therapy for his leukemia if he returned to the Ukraine.

"I called the administrators at my hospital in Boston, the Floating Hospital for Infants and Children of New England Medical Center, to see if there was any possibility of offering free care to this special child. They agreed, provided I could prove that his parents' income was below a certain level. As factory and construction workers with very low salaries, they easily qualified."

A month later, Vova's mother Olga arrived in Boston from Chernigov in the Ukraine to provide emotional support for her son and to give consent for the very aggressive salvage treatment

Molly was proposing. Vova's tearful mother had been effusive in thanking Molly for this second chance for her son.

The Ukrainian community in Cambridge surrounded Vova and Olga with all kinds of support, and the Children of Chernobyl Relief Fund provided financial and administrative help. But Vova was basically in Molly's hands. She finally got to see the child's wonderful smile once he got over his initial fears and learned to trust her.

"He keeps his stuffed bear Sasha within reach at all times," she reported. "He also can recite Hickory Dickory Dock in English. He must have learned it in kindergarten at home in Chernigov. He's a smart little guy with a passion for books, but when the chemotherapy gets to him, he likes to curl up with Sasha and watch 'Tom and Jerry' cartoons on TV.

"Vova and his mother have been in Boston for about eight months, and he's doing very well. However, I'm sure it will be more than a year before he's well enough to go home."

Molly offered workshop attendants copies of the chart of drug names and their Western equivalents she had written, as well as a list of possible alternatives to some of the drugs that might be mentioned in the histories of the children coming to the States for R & R. "I have to emphasize that the chart represents our best guesses in many cases, but if you think it might be useful, I'm happy to share it with any of you."

❊ ❊ ❊

MICHAEL WAS NEXT to speak. He described Citihope's six relief trips to deliver a million dollars' worth of medicine and supplies and the twenty-four children the organization was arranging to bring to the United States that summer. He was surprised at how favorably impressed the delegates were with this level of activity. Apparently Citihope, as small and grass roots as it was, could operate in the same league as major relief organizations such as AmériCares and Project Hope—at least as far as Chernobyl relief was concerned. And Citihope, he discovered, was the only organization focused exclusively on the needs of Byelorussian children at risk.

"I'm really interested to know if anyone has licked the bureaucratic and logistical nightmare involved in transporting medical cargo to the Soviet Union," he asked.

No one had an answer. He discovered that it was a problem every compassionate-response group had encountered to one degree or another and, they all agreed, it was one that surely would get worse as the political and economic situation in the USSR continued to deteriorate.

Michael also touched on the sticky situation of obtaining medical insurance for the children who would be coming to spend as much as eight weeks' time in the U.S. in the summer. "Our host organizations haven't been able to find a source of reasonably priced, short-term insurance. It looks like they are going to have to take the children uninsured and hope for the best. This is a serious risk for these small, grass-roots groups, and an accident or major medical crisis for any one of them could put an end to the entire program of bringing these children to the United States for rest and recreation."

Jerry Courtney of the YMCA responded. "Keep trying to find a source for the most basic insurance for the healthy children, but forget insurance for the diagnosed children. It's impossible. Instead, make advance medical arrangements—in writing—with cooperating hospitals and doctors."

Nearly every organization represented had encountered similar barriers in the course of its efforts; no one had a simple solution, but everyone was willing to pool resources to find the answers. The pace of the informal networking quickened.

World Vision's representative, Bob Mitchell, pledged financial help for the purchase, transportation, and distribution of medicine.

"Project Hope is more than willing to share space on the cargo planes it regularly sends to the Soviet Union," said Bill Walsh, the spokesman for that group. "And we have a staff of oncologists in Moscow and Minsk who could examine and screen your children before they come to the United States—if that would be helpful."

Walsh also hinted that President Bush was about to launch an initiative to ship medical supplies to the Soviet Union and that Project Hope expected to be the private, nonprofit partner in this effort. Then he touched on the need to address still another crisis in the USSR that extended far beyond its need for food and medicine.

"That country is experiencing a Chernobyl of the spirit, a deep and widespread despair over the tragedy that has befallen their land. We need to find ways to address that as well as the more obvious medical needs."

Each delegate had encountered that sense of despair in the families affected by Chernobyl. But it appeared that the kind of hope they could offer really did help. Knowing that people in the rest of the world cared and were working to provide help for their children seemed to loosen the grip that Chernobyl had on people's souls.

Bob Mitchell offered a response. "World Vision is particularly interested in providing mental health resources and counseling services to youth at risk in the contaminated regions. The need seems to be overwhelming."

Michael slipped into an empty chair next to Mitchell to discuss an idea he had. Less than a year later, World Vision had contracted with Michael and Citihope to launch just such a mental health program, in Gomel, in the "Chernobyl Crescent" of Byelorussia.

After lunch the participants got a chance to firm up some of the partnerships that had been formed during the morning session. For Michael, a natural networker, it was a feast of strategic resources. He didn't even have to pick up a phone; he just made the rounds and struck the deals that would benefit the children of Chernobyl and further each organization's interests.

Before the day was over, he had linked Citihope with Project Hope for transportation possibilities, nurtured World Vision as a partner in providing medicine and mental health services, and invited representatives of the YMCA and the American Cancer Society to go with him on his next delegation to Byelorussia. Michael also sought out Molly Schwenn and urged her to come with him in June and share her much-needed expertise as a pediatric oncologist.

As it turned out, the connection with Ed Baum would help provide some of the funding for Molly to make the trip, as well as accommodations for one of the children Citihope would bring to the United States for treatment.

By the time he boarded his flight back to San Francisco, Michael's spirits had been lifted so high that he hardly needed a plane to carry him home.

PART FOUR

Thin clouds drew a veil over a glaring June sun. A humid breeze rippled the meadow grass in the broad clearing. The sandy, pale-colored earth stood in random heaps of loam, the work of machines, unlike the rolling hillocks of the surrounding countryside molded by weather and time.

This was the buried village of Karpovichi in the Gomelskaya region of Byelorussia, less than twenty miles from Chernobyl's nuclear reactor No. 4. A set of rutted vehicle tracks led into the clearing and disappeared into the thick grass that covered all the evidence that people had once lived, borne children, farmed, and died here. Karpovichi was an empty, foreboding place—a void, a radioactive wound that would take tens of thousands of years to heal.

But there was also the crane—a huge bird nesting on a half-buried utility pole at the very edge of the clearing. She surveyed the grassy grave, stretched her expansive wings, and soared across the summer sky, gliding on the thermals that rose from the dead and buried village.

A crane—the same bird that had come to symbolize healing and peace after the nuclear holocaust at Hiroshima—had now come to nest above Karpovichi, one of Chernobyl's dead places. At Hiroshima a shrine recalls the folded paper cranes of Sadako, the child who died of her radiation wounds before she was able to fold the thousand origami cranes that legend said would restore her health. Visitors still place folded cranes at Sadako's memorial. At Karpovichi, a live crane soared over the buried wreckage of a dead village, promising rebirth and resurrection from the ashes.

12

"THE CAPTAIN GOES DOWN WITH THE SHIP"

"WELCOME, WELCOME, my friends," said Nikolai Markovsky as he embraced Michael Christensen in front of the mayor's office in Narovlia, a tidy, prosperous-looking city of ten thousand at the edge of Chernobyl's Dead Zone.

Michael waited for the eight other members of the Citihope delegation to climb off the chartered red bus, which had carried them seven hours from Minsk, before he explained, "This is Nikolai Markovsky, the former mayor of Narovlia. Now he's a national bureaucrat."

Michael had met Markovsky the previous Christmas, and this summer reunion was punctuated with smiles and embraces. "We've got a busload of medicine and medical supplies for the district hospital, and vitamins and baby formula for your children. Do you want us to unload it now?"

"Let's talk first," the small and wiry Markovsky replied as he directed the group into the unassuming white stucco building behind the statue of Lenin in the square. "We have prepared some refreshments for you and your friends."

Along with Michael and his coleader, Citihope radio producer Judy Ferguson, was Michelle Carter, who was making her first trip with the organization. With her she had brought more than forty cartons of the donated medical supplies that had filled the organ closet of the Congregational Church of Belmont.

"We've also brought a medical team on this trip for the first time," Michael said as he introduced Dr. Molly Schwenn, the pediatric oncologist from Boston; Dr. Andrew Davis, a Chicago internist who had visited Chernobyl accident victims in Moscow in June 1986; and Victoria Rakowski, an oncological nurse with the American Cancer Society in Michigan.

"Yes, you said in your fax that you would be bringing doctors," Markovsky replied as he led the group into his office, which looked remarkably like every other official's office Michael and Michelle had been in on this and previous trips. Unique to Markovsky's blond conference table were boxes of Birds' Milk Chocolate Candy, the local product for which Narovlia is famous throughout the Soviet Union.

"We have arranged for your doctors and the nurse to examine patients at the district hospital and consult with our doctors. Dr. Adam Nikonchuk, the chief doctor at the hospital, will be here shortly."

Michael finished by introducing Mary Beth Ormiston of the Michigan YMCA, whose organization had hosted a group of Byelorussian children with cancer at a camp in the summer of 1990; Asa Tribble, a member of the Hutterian Brethren community in upstate New York that would host a group of children in the coming summer; and John Leach, an urban ministry intern and graduate student from Oregon.

Markovsky sat down at the head of the table, under the now-familiar portrait of Lenin. He motioned to the interpreter to take the seat at his left and began talking as if he already knew all their questions.

"I used to be the mayor of this city, but now I am a member of the Union Committee on Chernobyl. I have come back to Narovlia to be your host and to help tell the story of this 'free-choice' city. We are about forty kilometers from Chernobyl, just outside the Dead Zone in the Gomelskaya district. This region is still contaminated by radiation from the explosion."

It was the "free-choice" designation, which allowed residents to decide for themselves whether to stay or leave, that created headaches for Markovsky, an intense, dark-eyed man who was given to breaking into broad smiles that revealed more than one gold tooth.

"The state directive was simple for residents within the thirty-kilometer Dead Zone around the reactor—one hundred percent evacuation. Everyone was ordered to leave," Markovsky explained. Within thirty days after the explosion, the militia moved in with trucks and buses, and residents were packed off to relocation sites with only the belongings they could carry in their arms.

"But the residents of Narovlia have always been free to choose whether to stay or go," he said. Choice, however, implied decisions based on clear and ample information, and Markovsky was the first to admit that the biggest problem in the days after April 26, 1986, was a lack of such critical information.

Rumors had been flying that something horrible had happened at the nuclear energy plant just over the border in the Ukraine, but party officials had remained officially mum—even while they had made private arrangements to speed their own families to safety outside the region. "I drove with my family and friends to the countryside to spend the day in the sun," Markovsky recalled. "We knew absolutely nothing until Monday, when we heard reports from the workers at the plant who had seen with their own eyes the black cloud and the burning fragments.

"The first evacuation order didn't come until ten days after the disaster. Before that, every communication said to maintain order and reassure the people. So I did that, and then we got the order to evacuate the villages in the thirty-kilometer zone. The telegram said, 'There has been an accident. Situation very dangerous for the people. Evacuate.' "

"Today," Markovsky said, "the situation is not so much better"—despite some highly publicized official orders. In October 1989, the Supreme Soviet had adopted a five-year plan and allocated seven hundred million rubles per year to resettle up to one hundred thousand people from more than four hundred communities. Very little of that money and virtually none of the relocation services had ever arrived in Byelorussia, and local frustration with the central government's inaction had spilled over. The republic had issued its own international appeal for help in relocating two million people, one-fifth of its entire population, by the end of 1991.

Slowly, the central government had awakened to the fact that its lack of response to the crisis was becoming a public relations nightmare. So in April 1990 the Supreme Soviet updated its program "to liquidate the consequences of Chernobyl." The new order recognized that earlier efforts had been "inadequate" and that any effective effort would have to coordinate long-term relief for Byelorussia, the Ukraine, and the Bryansk region of

Russia. The primary goal would be to resettle families, especially those with pregnant women and school children.

Markovsky described the order as "most welcome," but later in the trip Michelle and Michael talked to local members of the Byelorussian Popular Front who had nothing but contempt for the promises of the central government, as well as the Communist officials in Minsk.

"These are only words," one of them said as the Americans sat around his kitchen table. "There has been no action. Families have been on lists to be resettled for months, even years, and they never get to the top. Some have even been given resettlement dates, but the dates come and go and nothing happens.

"You should not listen to the mayor. He's a traitor who knew everything about the danger and told no one."

But that morning at the conference table, Markovsky maintained that he had been just as much in the dark as everyone else. Even after the dangers became known, he lacked the authority to tell people to stay or to leave. They had to choose for themselves.

"About two thousand residents have already left," he said, "and most of them left because they feared for the health of their children. I understand their fear. I have two children, eleven and fourteen, still living in Narovlia. They have been subjected to high levels of background radiation every day since the accident and they eat contaminated food every day, too, just like all the children."

"Will you leave Narovlia?" Michael asked.

He shook his head. "The captain goes down with the ship."

"What about your children?"

Markovsky shrugged, "They know it's my job."

He went on to explain that none of the many radiation experts—Soviet or Western—who had converged on the area with dosimeters and monitoring equipment had provided much clear information either.

After a period of belabored disinterest, Soviet scientists had launched their own studies, which reported the likelihood of genetic and fetal abnormalities and increased incidences of leukemia and thyroid cancer in children living in the contaminated zones. However, it will take decades to establish significant statistical patterns and linkages.

Then, in June 1991, the International Chernobyl Project of the International Atomic Energy Agency (IAEA) issued its

"Vienna report." It poured cold water on the findings of the Soviet scientists and described the radiation anxiety in the region as "wholly disproportionate" to the radiation contamination measured.

"So, there are different points of view, especially among doctors and scientists," Markovsky said. The doctors who lived and worked in Narovlia, however, appeared to have made up their own minds. "Our hospital is short eight doctors now and seven more are leaving in July.

"Most people believe the health risk is far worse than the very conservative report of the IAEA. The doctors themselves are leaving, the doctors who are concerned about the health of their own children, and other specialists and qualified young people."

Dr. Adam Nikonchuk, the chief doctor of the local hospital, joined Markovsky in the discussion at the conference table. "Local doctors are very aware of the consequences of radiation now, but years ago no one knew. Thyroid cancer in children didn't exist here before, and now it does. Now they know—and many are leaving."

But leaving wasn't as simple as it sounded. "It can take up to eighteen months," Markovsky explained, "for a family to be relocated outside the contaminated region. For some, even longer. Every effort is made to move people to an area where they have relatives, but that doesn't always happen.

"Soviet law also requires enterprises in the relocation areas to provide jobs for the resettlers, but jobs are scarce and homes are even harder to find."

Some provisions were made for those residents of Narovlia who had chosen to stay. The state offered significant compensation. "It includes a fifty to seventy-five percent increase in salary, food imported from outside the region, and forty-two days of holiday a year away from the area," Markovsky said. ("Coffin rubles! Grave money!" scoffed the Popular Front members who later spoke to Michael and Michelle.)

However, there was no disagreement that it was important for children to have periods of rest and recreation outside the contaminated region. Markovsky was working with the Byelorussian Children's Fund to send some children, including his own eleven-year-old daughter, to America for part of the summer of '91 under the auspices of Citihope.

To deal with the problem of food grown in irradiated soil in the area, the city established inspection sites where residents could bring produce from their own gardens to be checked. Markovsky said that milk from local dairies had to meet radiation standards more severe than those in the United States, so that the already elevated radiation levels in the area would be taken into account.

Contaminated milk was one of the tangible issues that Markovsky and other officials addressed in practical fashion, but there was little they could do to dilute the anxiety and stress that life in the region created in the residents. He said that eating disorders were turning up as some people stopped eating altogether rather than consume contaminated food. Children were afraid and would no longer eat the fruit and vegetables available to them. Their thyroid glands and immune systems were doubly affected—by radioactive contamination and by poor diet. Every health problem, every sniffle, every cough was perceived to be a result of the accident.

After the meeting, the medical team of Molly, Andy, and Vickie split off to go directly to the district hospital with Dr. Nikonchuk to meet with patients the staff had selected for them to examine. The others boarded a small bus with four-wheel drive and oversized tires for a journey into the Dead Zone.

❉ ❉ ❉

AS THEY BUMPED along rutted, overgrown dirt roads, the Americans remembered the media images they all had seen of the ghost villages, with weed-filled gardens, gates hanging from one hinge, breezes rippling pages of books abandoned at the last minute. They were prepared to find the broken dolls, broken windows, and other remnants of dreams left behind by villagers ordered to take only what they could carry and, in thirty minutes, leave forever the only homes most of them had ever known.

Then they saw a haywagon. In the evacuated village of Gridney in the total-exclusion zone, about ten miles from Chernobyl, was a scene right out of a Constable landscape. The haywagon, pulled by a well-fed and brushed chestnut mare with a nervous colt trailing at her side, was driven by a wiry, rosy-cheeked farmer in his sixties, with his plump, kerchiefed wife perched on top of the hay.

"Stop, stop the bus!" Michelle shouted to the driver in a confusing mix of Russian and English. "I want to talk to them."

The farmer and his wife smiled, first cautiously and then broadly, as the Americans piled out of the bus with cameras and questions. The commotion they created brought more curious villagers, including a young woman with a baby in her arms. Everyone shook hands and the Americans pulled out photographs and postcards to tell their stories without the benefit of words.

Michelle was determined to interview the farm couple. She started out with questions in Russian, but their replies were too fast and too lengthy for her ability in the language, and she had to pull the interpreter into the conversation.

Were they sick? "Do I look sick?" was the farmer's robust reply. He did not.

Were they worried about living inside the Dead Zone? "I have food to eat and a house to live in with a good wife. Why should I worry?"

The farmer and his wife had stayed away a year after they were evacuated. When they returned, they found that the militia had taken over their farmhouse. "We couldn't get them out. And when they did leave, they took our things. It was a mess," the wife called out from her seat on the hay. "But now everything is good."

They said they had children and grandchildren living in Narovlia, but they chose to make their life in this officially abandoned village and to eat the produce of their field and their livestock. They had not made any philosophical decision to do daily battle against the effects of constant, low-dose radiation; they had just gone home.

Yet there *had* been desolation in this farm village of Gridney. The militia had come in the early summer days of 1986 and packed one hundred and fifty families into buses and trucks, with their possessions tied into blanket bundles. Some had left willingly in fear and confusion; others had to be pulled from their land.

Now, five years later, thirty families had come back. Most (but not all) of them were older and without children. They had weighed the danger of something they could neither see nor touch against a dependent, unfamiliar life and had chosen to return.

The Americans said their goodbyes, handed out Polaroid snapshots that they had just taken of the residents, and climbed back onto the bus.

Just a few kilometers down the road from Gridney, the group passed another village whose residents had not come back. It was a newer community, with sturdy, concrete-block houses and a two-story secondary school. There the Americans saw abandoned books and dolls and plates on the table of a house on *Uleetza Novaya* (New Street). Through the rubble on the floor, the face of Vladimir Ilyich Lenin gazed up with an enigmatic smile from the cover of a youngster's history book.

❋ ❋ ❋

NOT A TRACE remained of the village of Karpovichi, a little further down the road. Background radiation levels were so high in this "hot spot" that officials bulldozed the houses, barns, and shops into the sandy Byelorussian earth so there would be nothing drawing villagers back into the region. It was one of four villages in the contaminated zone that had been totally buried.

It was here, in Karpovichi, that the Americans caught sight of the nesting crane in the distance. With cameras poised, they moved as quietly as nine adults were able, to within twenty-five yards or so of the nest. As they began to inch closer, the crane lifted herself to the edge, extended her wings to full span, and swooped off over the heads of the crouching intruders. As the Americans pondered the wonder of a crane in flight over the buried village of Karpovichi, Michael recalled the Byelorussian legend that Vladimir Lipsky had told him during his first trip to the region: "When cranes return, people will soon follow."

Would that be true of the buried village of Karpovichi? "Perhaps," Markovsky said, "in a thousand years."

The next stop wasn't nearly so poetic.

"It looks like a construction site for a small city," Michelle told Mary Beth as the bus approached. "But what on earth is that smell!"

In a muddy wedge of scarred earth, workmen, some with handkerchiefs tied over their noses, were using heavy equipment to dig deep trenches. In some of the older trenches, huge, concrete sarcophagi had been constructed.

"What are they building here?" Michael asked. "What's in those concrete tanks that smells so awful?"

"This is the burial site for meat that has too much radiation, that can't be eaten," explained one of the district officials who had accompanied the Americans. "The rejected meat is dumped into the concrete boxes, and then maggots are allowed to eat the meat until each box is filled.

"Then it's sealed with a concrete lid and buried under ten meters of soil. More than three hundred tons of contaminated meat has been disposed of this way since the accident."

The workers, who toiled in the summer heat and the stench of the rotting meat, were rotated in and out on a weekly basis. During their week on duty, they ate in a common *stolovaya*, or dining room, which would have been inviting in another setting. They shared a dormitory with a friendly yellow dog of indistinct breed who loved company and registered a shockingly high millirad-per-hour count when one of the Americans held a geiger counter close to his wagging tail.

✳ ✳ ✳

ALONG THE ROAD back to Narovlia, the bus stopped at an untended cemetery with a fresh grave marked by a huge pine cross. On the bus ride out, the funeral had been in progress as mourners carried the rough casket up the hillside, followed by a man dragging a cross, much like Christ on Calvary. Now that the crowd had left, the Americans wanted to stop and linger.

The overgrown graves of significant members of the community occupied the high ground, but the new grave was just a few yards from the road. The funeral had honored a seventy-one-year-old woman who had lived most of her life in this village and whose family had been allowed into the Dead Zone to bring her home for burial.

In the freshly turned earth of that grave, about fifteen miles from reactor No. 4, Michael's ever-present radiation meter recorded the highest ground radiation level the group would see anywhere in the region—six hundred millirads per hour, about sixty times the normal background radiation in an American city.

The rad count was shocking, but it was the swarms of ordinary mosquitoes in the tall grass of the cemetery that drove the group back into the bus.

Michael and Michelle both pressed Markovsky to take them to the reactor itself. Ever the journalist, Michelle had been working every possible angle to get to Chernobyl, but "not possible" was the standard reply. She was particularly reluctant to accept the official denials since the local television cameraman who was videotaping their visit had been there with a group of Australian journalists just the week before. She sensed, perhaps unfairly, that officials felt the need to protect this female journalist in a way that they wouldn't protect male journalists.

Yet the group did see the reactor—from a distance. The bus passed through one last barbed wire barrier (after three militiamen unlocked it and lifted it out of the way) and continued down a tree-shaded country lane with elegant and substantial (and quite obviously abandoned) *dachas*, summer cottages, on the banks of a meandering creek. Finally, the bus came to a stop in a small clearing just above the Pripyat River.

The dominant feature in the clearing was a ramshackle, wooden tower about twenty-five feet high. From the landing at the top it was possible to see the massive buildings that made up the four nuclear reactors on a rise above the Pripyat. The tower was about two miles from the power plant, surrounded by the verdant riverscape that belied its three-digit millirad readings. But the message of the barbed wire and the "forbidden" signs was unmistakable. That dark, cement-gray tomb in the distance was still leaking death.

Each member of the delegation, even seventy-year-old Asa, climbed the tower for the chance to see the source of so much pain and destruction.

The bus ride back was quiet. By then the senses of most members of the group were on overload, and Michael kept hearing Markovsky's haunting words: "The captain (and his family) goes down with the ship."

✻ ✻ ✻

WHILE MOST of the group had been driving from one militia checkpoint to the next through the thirty-kilometer zone, the medical team had been meeting with Dr. Nikonchuk—a large, beefy man with an unassuming, no-nonsense tone—and his staff at the district hospital.

Dr. Andrew Davis, a physician with a bent for academic precision and careful note-taking, asked about those disappearing doctors that Markovsky had mentioned. Nikonchuk, a general surgeon for twenty-eight years, said the normal complement of physicians for his two-hundred-and-ten-bed hospital was fifty-two. That number had fallen to twenty-four in 1990, and seven more would leave the following month.

"The worst part," he said, "is that we will lose key specialties such as anesthesiology, endocrinology, and dermatology when the next group leaves."

Andy had already observed that doctors were numerous in the Soviet Union, but they typically completed their training by age twenty-four or twenty-five (compared to about thirty in the United States). Outdated texts and materials were frequently used. Low salaries for doctors, who are service-providers, not producers (the highest class in the communist value system), reflected their lack of status. Andy learned that a doctor typically earned about two hundred rubles a month, compared with an average of three hundred rubles for a factory worker.

"More talented and ambitious practitioners appear to be seeking employment elsewhere in the country," Andy wrote in his notebook, *"and several asked us about further training in the West."*

By the fall of 1991, many of the most talented doctors were taking advantage of the new entrepreneurial climate to open private practices and buy their pharmaceuticals on the black market from the newly emerging *mafiya*. Now ordinary citizens could have the quality health care that had previously been available only to the *apparatchik*, as long as they had plenty of the new measure of clout—hard currency.

For many it was the only way to get even the simplest of medications. The shelves of the drugstores were as empty as those in the grocery stores. It was nearly impossible to buy aspirin, let alone antibiotics. More sophisticated pharmacy items, such as feminine hygiene products, were nonexistent. The American women who traveled with Citihope routinely left packages of tampons in their hotel rooms when they departed, as gifts for the hotel staff. One *dzhernaya* at the Hotel Pripyat ran after Michelle as she was leaving, to thank her for the gift.

Andy continued his detailed note-taking: *"We also got a sense for the uneven supply and quality of Western aid. We encountered medical supplies from West Germany, France, Switzerland, and Japan, but difficulties arose from incompatibility and unfamiliarity.*

"Soviet doctors are also somewhat cynical about the outdated medicines brought in by some relief agencies, with flourish and favorable press in the originating country. There is also clearly some jealousy between in-

stitutions and very real concerns about the siphoning off of children of Chernobyl aid to clinics in other parts of the country."

Andy carried copies of the International Atomic Energy Agency report with him and asked doctors their opinion of its findings. *"Across the political spectrum, we found unanimous disagreement with this cautious dismissal of the issue of radiation-induced illness in the contaminated regions. Most thought that the agency inspectors' brief stay and technocratic focus had blinded the group to the obvious illnesses confronting seasoned doctors every day."*

The medical team—Andy, Molly, and Vickie—was exhausted by the time the rest of the Americans met them at the district hospital. They had been with patients or staff members constantly since mid-morning. While the three of them drank tea at a much-delayed reception, the others unloaded the thousands of pounds of vitamins, antibiotics, and surgical equipment the Americans had brought with them from Minsk.

As the boxes appeared, Michael and his coleader Judy Ferguson exchanged knowing glances, recalling the ordeal that once again had been involved in getting that desperately needed cargo to its destination.

The drama that had been played out at the Aeroflot cargo desk at Kennedy International in New York had been like all the earlier relief trips. UPS had delivered the load to the airport, and Sharon Moore, who was not making the trip this time, nonetheless had to reassume her bulldog role in an attempt to get the Soviet airline to take the shipment without charge. After much effort, Aeroflot finally agreed to send everything the next day "at minimum charge," to be paid later.

And so the cargo had arrived—not the next day, but three days later. Paul Moore, who was getting quite proficient at shepherding medical supplies, accompanied it to Minsk without further problems.

In Narovlia, the entire medical staff at the district hospital applauded each box as it was carried in. The head nurse opened one of the cartons of "rescued" surgical equipment that nurses on the San Francisco Peninsula had donated. Wearing a white, starched toque that looked like a chef's hat, she clapped her hands to her face and shrieked. Through her tears, she embraced each of the Americans, offering the only words she knew in English: "Thank you, thank you!"

She uncovered the different-sized scissors, scalpels, tweezers, and tongs with awe and then waited patiently for an explanation that the small, sealed packets held disposable scalpel blades. Michelle concentrated carefully so she could play this scene back to the nurses at home. They needed to know that, to the head nurse at the district hospital in Narovlia, their discarded surgical supplies were a gift beyond price.

13

"WHAT DOES THE LORD REQUIRE OF YOU?"

IT WAS NEARLY midnight the following day when the bus pulled up to the portico of the Hotel Belarus. As the sleepy Americans dragged themselves out of their seats, Michael reminded them that breakfast would be early the next morning, so they would need to gather even earlier for morning devotions.

Michelle was perhaps the only one who didn't groan. An early riser who was used to predawn wake-up calls, she would enjoy the opportunity to start the day with coffee, conversation, debriefing, and a few minutes of worship. She found morning devotions to be a useful way of reminding herself, and the others in the group, of the motivation behind their mission to the Soviet Union. They were not just curious visitors, but rather Christians of various backgrounds and traditions on a mission of mercy and good will.

This morning it was Michelle's turn to prepare the devotions, and she knew what she would do. A verse from the Old Testament, Micah 6:8, had been the theme in November for the last United Church of Christ delegation, and it seemed even more appropriate to this trip. Once the group was settled on the couch, chairs, and floor, she took out her Bible and read:

> . . . and what does the Lord require of you
> but to do justice,
> and to love kindness,
> and to walk humbly with your God?

What single Bible verse could more aptly capture what these ten Americans were about on this visit?

"So how do we 'do justice' here among the children of Chernobyl?" she began. "I don't think God intended for us to be instruments of justice in the legal sense, but a sense of divine righteousness as well as 'rightness' is required. We must behave 'rightly' and 'justly' among whatever people we find ourselves.

"As for 'kindness,' other translations use the word 'mercy.' I think God is asking us to give the most pure form of kindness and mercy, which is love without pity or judgment.

"Then comes the hardest part for most of us, 'walking humbly with our God.' Few of us are into real humility, and it is even harder for us to 'walk humbly' when we find ourselves in the position of 'giver' and 'provider' to those in need. This perhaps is our real test. According to Micah, God requires that we be humbled, that we humble ourselves, in fact, before we can truly give of ourselves."

AFTER BREAKFAST, the red Intourist bus hired by the Byelorussian Children's Fund took the group to the three-year-old Anti-Chernobyl Diagnostic Center, the cornerstone of Byelorussia's effort to establish scientific benchmarks with which to measure the health effects of the catastrophe on the eight hundred thousand children at risk in the republic.

Dr. Larissa Sivolobova led the visitors through a building dedicated to monitoring every possible aspect of the health of eighteen thousand children who live in the most seriously contaminated regions. The children came twice a year for examinations that would create a bank of hard data. Soviet scientists needed evidence to disprove the recent IAEA claims that the explosion had not caused any serious health problems.

At the end of the tour, the Americans pressed the willowy, redheaded Dr. Sivolobova for statistics on what the doctors of the center had discovered in the first three years of operation. She refused to be nailed down on specific numbers of cases, but she did say that eighty-eight percent of the children seen at the center had some physical or psychological impairment.

About six weeks later, on August 16, an article in *The European* reported that Byelorussia's hard-line premier Vyacheslav Kebich had lashed out against the Kremlin's mismanagement of Chernobyl aid. Kebich quoted a forecast by Valentin Kondrashenko, the chief psychiatrist at the Byelorussian health ministry, that the republic "could become a land of mental defectives and psychological cripples within two or three generations."

Kondrashenko based his comments on the results of a psychiatric survey, which found that nearly half the children under fourteen in the contaminated provinces of Gomel and Mogilev "showed brain abnormalities affecting their reasoning capacity and psychological well-being." He joined the list of scientists and politicians asserting that Byelorussia was "suffering genocide by radiation."

Up to that time, only the Popular Front had been making such statements. Now even the Communist Kebich was charging that desperately needed Western aid for Chernobyl victims, both medical and monetary, "was being diverted by Moscow officials to line their own pockets."

The official denials of health risks, the article said, were convincing no one in Byelorussia:

> Whatever the experts say, the Byelorussians have made their own decision. Population growth is down from 7.4 to 5.4 per thousand. Many young couples, fearing genetic damage, have decided not to have children. According to Foreign Minister Pyotr Kravchanka, this [drop] alone poses major demographic problems for the future. Birth defects, he told a United Nations meeting in Geneva last month, have risen from 4.1 to 7.7 per thousand live births—very close to the 8 per thousand threshold figure which would threaten the genetic fund of the Byelorussian nation.

❋ ❋ ❋

AT THE ONCOLOGICAL INSTITUTE, the Americans met Dr. Reiman Ismail-Zade, dark, handsome, and eager to help, who would be bringing ten of his cancer patients to Mary Beth's YMCA camp in Michigan later that summer. His hospital was significantly less well-equipped than Dr. Olga's Hematological Center, but colorful murals brightened what would otherwise have been dreary rooms painted "institutional green."

Here the children's afflictions were more obvious. Two teenage boys, each missing a leg, rested on crutches at the edge of the group of Americans. When a member of the group asked how long it would be before the boys could be fitted with prostheses, she was told that artificial limbs were very difficult to obtain in the Soviet Union and that the boys surely would not be fitted until they had stopped growing. The idea that they should have a series of progressively larger prostheses while they continued to grow was quite out of the question.

While Judy Ferguson and John Leach got out the Polaroid and took pictures of the children lining the halls and spilling into the common room, Michelle wandered into one of the rooms. It was empty except for a frail, whisper-thin, and obviously uncomfortable teenage girl and her mother. The girl said her name was Tanya and she was thirteen. Her mother, Irina, pulled a wheelchair up to her bed and asked if she wanted to go out with the other children to get her picture taken. Not yet, Tanya said; in a minute or two she would.

Later, as the group was handing out stuffed animals from their huge duffle bag of goodies, Irina appeared in the hall, pushing the wheelchair in which Tanya, in a red velour jogging suit, sat bent over. Judy turned to take a picture of Tanya and her mother, and she asked Irina to lean over next to her daughter so they could be photographed together. When Tanya was handed the snapshot, she smiled and passed it to her mother, who could no longer contain her tears. She turned and wheeled her daughter back to her room.

A little while later, Irina stood on the fringes of the noisy group. Michelle asked if she would allow an interview. She seemed surprised that anyone would be interested in her story, but agreed. Michelle and an interpreter sat at a table in the common room and Irina talked.

"Eighteen months ago, Tanya complained about a pain in one of her legs." She spoke so quietly that both the interpreter and Michelle had to strain to hear her. "I told her she was growing too fast and the pain would certainly go away very soon. But it didn't go away. It only got worse. Finally I took a day off from my job at the factory and took Tanya to a clinic where they decided to do some tests."

In a few weeks, the teenager was admitted to the Oncological Institute, where surgery was performed for a neuroblastoma of the spine. Despite the debilitating effects of radical surgery and the chemotherapy that followed, Tanya seemed to improve dramatically, and eventually she went home.

Six months later the pain returned, along with growing weakness in her legs. Tanya was brought back to the institute for radiation therapy, and she had been there ever since. Now her condition was rapidly deteriorating. "She hasn't been able to walk for several weeks," Irina said. Tears ran down her face, but she made no attempt to wipe them away. It was as if she was too tired of Tanya's pain, too tired of her own tears, to bother.

"Now she can't even sit up without help, and she's in terrible, terrible pain all the time. I think she's going to die very soon." Then Irina reached inside her blouse and pulled out a cross on a chain around her neck.

"I'm a believer and Tanya is too. I saw the priest with you. Would he bless my daughter?"

Irina had noticed the cross Michael wore around his neck and identified him as a clergyman—even though he wasn't wearing a cleric's collar. Michelle found him and asked him to pray with Irina. Together they went into her room, where Michael first took out a vial of sacramental oil and anointed Tanya's forehead in the sign of the cross. With his hands on her head, he prayed for healing, deliverance, and an end to her suffering.

When Irina came out of the room, Michelle put her arm around her slumped shoulders and asked if she had other children.

"I have a son Sergei who's ten. I haven't seen him in a long time. My husband looks after him the best he can, but mostly he gets along by himself. I worry about him, but he has to be strong because I must be here."

Irina said she hadn't been to her job for many months, but she continued to draw her salary. "We couldn't live without it. I will go back to work when I can. But that's another life. My life now is here, always with Tanya, until she doesn't need me any more."

THE GROUP WAS climbing into the bus outside the Oncological Institute when someone noticed that Asa Tribble was not there. Judy and John went back inside, up to the second floor, and poked their heads into room after room until they found seventy-year-old Asa, dressed in his familiar black pants and vest and his long-sleeved white shirt, sitting on a child-sized cane-back chair. He was surrounded by the entire family—father, mother, brothers, and sisters—of one of the patients. He had a plate of nuts on his lap that the family had given him, and they were all chattering away as though they had found their own dear *dyedushka* (grandpapa) and brought him into the family circle.

Asa had taught himself some Russian simply as an exercise in discipline, and when his community of Hutterian Brethren decided to send a representative with the Citihope group, they had chosen him. The choice was inspired. His simplicity and all-accepting love were obvious to the children he met.

The sight of white-haired Asa huddled with a giggling child, flipping the pages of his palm-sized Russian-English dictionary, became an enduring image for the other members of the group. There was a gentle humility about him that drew strangers to his side. Vladimir Lipsky had noticed it earlier when he said, "You can see in that man's eyes the lightness of his soul."

❋ ❋ ❋

THE FIREMEN'S HOSPITAL got its name because a number of the "liquidators," the men who fought the initial reactor blaze, were treated there afterward, and doctors there continued to monitor the health of those who survived. But endocrinology was the primary discipline practiced at the hospital, which was officially known as the Radiation Science Institute.

It was located several miles outside of Minsk, beyond a patch of the densely forested land that covers two-thirds of the Byelorussian landscape. The Intourist bus followed a well-maintained road through the canopy of fir trees that finally opened up to the park-like grounds of what had been a sanatorium for the *apparatchik* and their families.

It was instantly clear that this had been no ordinary clinic. The broad expanse of the entrance foyer opened to a reception hall with floor-to-ceiling windows on two sides and an eight-foot marble bas-relief of wood nymphs and maidens filling the third wall. Exquisite wood parquetry in intricate patterns covered the floors. Compared to the crumbling institutional decor of the Hematological Center and the Oncological Institute, the Firemen's Hospital was a most inviting place.

The group was met by Dr. Slavomir Khinevich, an endocrinologist in a white laboratory coat and the tall, white hat that never failed to remind the Americans of a chef. A tall, sandy-haired man of about forty who spoke some English and understood far more, he was particularly eager to help the Americans understand the impact of the radiation of Chernobyl on the thyroid glands of the children he treated there.

Andy and Molly pressed him about the incidence of thyroid cancer in children before Chernobyl. "Almost none. Perhaps I saw one or two cases, but now this hospital is full," he said. Immediate surgery was the treatment of choice for childhood thyroid cancer, followed by radiation in some cases.

He admitted that some of the parents were concerned about the use of radiation therapy as treatment for cancer that they were convinced resulted from exposure to the radiation of Chernobyl. "But we talk to them and they trust us." He added that early diagnosis and treatment, as well as lifelong hormone therapy, could offer these children fairly normal lives.

Dr. Khinevich led the Americans along one of the sunlit halls with windows opening to the gardens below on one side and hospital rooms on the other. A number of children, from about six to twelve years old, popped out as Judy and John pulled out the Polaroid and bag of stuffed animals. Most of the children had "Chernobyl smiles"—the thin, white, two-or three-inch scars at the bases of their throats that were sure testaments to their thyroid surgeries.

An outbreak of chicken pox had put two other wards in quarantine that day, but the head nurse promised to deliver the toys and candy the Americans had brought for those patients.

A gaggle of older children followed the Americans downstairs, and John Leach pulled out the long, skinny balloons he had learned to shape into animals. As John strained to inflate the balloons and form the animals, a teenager named Misha tried out some high school English on Asa and John. While the other children were trying to guess the animals John was making, Misha struggled to name them in English. John's misshapen animals and Misha's broken-English names for them brought the excursion to the Firemen's Hospital to a noisy conclusion.

❈ ❈ ❈

BACK AT THE HOTEL Belarus, someone was waiting for Molly Schwenn. As she climbed off the bus, a tall man loaded down with packages approached her. "I am Vova's father. I have come to thank you for saving my son's life."

Alexander Malofienko was the father of the six-year-old boy Molly had been treating at Boston Floating Hospital. He had traveled all night by train from his home in Chernigov, the Ukraine, to meet Molly. He gave her presents to take back to his wife, Olga, and his son, and then ceremoniously presented her with a handmade clock. A disciplined doctor who succeeded in keeping her emotions in check most of the time, Molly was clearly moved by his expression of gratitude.

Molly had something for him as well. She reached into her backpack and pulled out a recent photograph of Olga and Vova. As he studied it, his eyes welled up. "Another year?" he asked through the interpreter.

"Just one more year and then they both will come home. Vova's really doing very well and he's a wonderful boy," Molly answered.

Later, Molly carried her clock and the presents for Vova and his mother to the elevator. "Vova's had a year of intensive chemotherapy drugs, with radiation to the brain and spinal cord," she told Mary Beth. "Next year he will get a less intensive program of maintenance chemotherapy. That won't be so hard on him." Looking down at the gifts, she added, "His father misses them both very much."

A MONTH LATER, Vova and Molly would go back to the Hole-in-the-Wall Gang Camp, where they had met the summer before. This time he would be the only child of Chernobyl there and would know enough English to get along well. Molly would review his treatment, with an eye toward sending him back to his father the following summer—free of leukemia.

❉ ❉ ❉

NEXT STOP would be the Children's Hematological Center, Michelle's first visit since November and Michael's first since January. The fruits of Dr. Olga's labors were everywhere evident. Computerized equipment filled the formerly empty laboratories and storerooms. Children walked up and down the halls and into a brightly painted common room, trailing mobile IV poles from which hung computerized IV monitors. A Western-style washer and dryer were being unpacked in an anteroom, and meals were dispensed from a common kitchen rather than from bags parents brought for their children. The walls of the patient's rooms had been freshly painted, and children's artwork and murals lined the hallways.

The new equipment and the facelift had produced a dramatic transformation among the patients and staff. Despite the obvious despair of children with life-threatening diseases, the nurses looked involved and interested instead of overworked and unable to cope. The children were out of their rooms and interacting instead of sitting on their beds tethered to stationary IV poles.

One of the most noticeable additions was the presence of Russian Orthodox nuns in dark habits, talking to the children, passing out art supplies, and comforting the mothers.

All the new equipment had been a gift from people in Europe, Japan, and the United States. On this trip, the Americans unloaded boxes of catheters, vitamins, sterile syringes, surgical dressings, antibiotics, immunoglobulin, and disposable "couplers" needed to use a blood purification system acquired from another donor. There was no methotrexate included with this shipment, and no need for it. The April delegation from Citihope had delivered enough to meet the hospital's needs for several months to come.

Now Dr. Olga was looking farther ahead. She told Molly Schwenn that architects had begun work on her dream—a children's cancer treatment center in Minsk. She pulled out drawings of a center that would incorporate all the Western medical treatment protocols for all forms of cancer, including bone-marrow transplantation. The financing would come entirely from donations being collected in Germany.

In spite of all this good news, Dr. Olga's children still had serious needs. She said her treatment center was doing very well in dealing with the initial onset of leukemia, but "it's not good for the children who come back to us a second time."

She also had to deal with an uncertain blood supply. At that time, nearly two-thirds of the children transfused at the Hematological Center showed symptoms of hepatitis. Dr. Nikonchuk at the district hospital in Narovlia was equally worried about spread of the AIDS virus during surgery. There was a real danger of transmitting HIV through inoculations with unsterilized needles, and Soviet HIV diagnostic kits—when they were available—performed poorly.

"Now, more and more, I am turning to parents and brothers and sisters to be (blood) donors for my children," she said. "But the real answer will be a state-of-the-art blood bank, which will be part of my new medical center."

Because Dr. Olga was now able to consider the very real possibility of a new, multidisciplinary treatment center, other medical professionals had begun to snipe: "Dr. Olga gets everything; we need medicine and equipment just as much as she."

That complaint was heard more than once as the Americans retrieved their cartons of medical supplies from a warehouse near the Byelorussian Children's Fund office. Michelle wondered if Dr. Olga had thought of sharing her communications and "marketing" skills with others who were treating children of Chernobyl.

Michelle also smiled when she heard Dr. Olga ask if the Americans had brought any "VITT-a-mins" for her children. "Did you hear that, Michael? Someone has been able to convince the good doctor that, even if they aren't a substitute for methotrexate, vitamins can be of real benefit in boosting the weakened immune systems of her children."

❋ ❋ ❋

BEFORE SHE LEFT the Hematological Center, Michelle had one last task. She needed to present a special gift that she had brought all the way from San Mateo. A friend who had heard about her mission of mercy had entrusted her with a digital wristwatch with a Mickey Mouse head—big, black ears and all—that stood up off the wrist. If the hands were squeezed, that famous Mickey voice would say: "Oh, boy! It's two o'clock!" (or whatever the hour might be).

Mickey Mouse had become something of a mascot for the trip, ever since Michelle had spent nearly an hour on the night train trying to reset the watch to Minsk time—without benefit of any instructions, since her friend had neglected to include them. She was relieved when she finally figured out how to poke the point of a ballpoint pen into a hole on the back of the watch and trigger the "set" mechanism. However, she hadn't realized that Mickey had other attributes besides telling the time when he was squeezed.

The first night in Minsk, after Michelle and roommate Mary Beth Ormiston had fallen exhausted into their beds at the Hotel Belarus, both women were startled awake by Mickey's shrill voice, saying: "Oh, boy! It's 1:05 A.M. Time to get up!" It took several seconds before Michelle realized that the voice was coming from the watch, tucked away in her suitcase. Mickey had gone on to report 1:06 A.M., and then 1:07 A.M., before she dug out the watch, found a pen to stick in the hole, and shut it off.

Michelle and Mary Beth had the rest of group breaking up with laughter the next morning as they told of their middle-of-the-night encounter with Mickey Mouse. It wasn't nearly as funny the following morning, when Mickey again chirped, "Oh, boy! It's 1:05 A.M. Time to get up!" Mickey's alarm was finally silenced the next day, after concentrated efforts by Andy, Michael, and Michelle.

But the watch delighted the children in each hospital the Americans visited. Michelle would crouch next to a child, ask, *"Vwee znayeti* Mickey Mouse?"* (Do you know Mickey Mouse?), and invite him or her to squeeze his hands. *"On govareet chas po-angleesky"* (He tells the hour in English).

Mickey made friends wherever he went, and Michelle intended to give him to one special child before she left Minsk. At the Hematological Center, one little girl, Anna, had followed her from room to room the whole time she was there. She had a bright, round, Slavic face with short, straight blonde hair. She told Michelle that she was eight years old and would be going home soon.

Anna looked quite healthy, and Michelle assumed that she was through with her treatment. However, when Anna took Michelle to her room and introduced her mother, she was told that Anna had relapsed with leukemia after two years in remission. If she went home, it would be because nothing more could be done.

Anna led Michelle to her bed and there, on the wall above her pillow, was a cutout of Mickey and Minnie Mouse. Mickey had found a home!

Michelle took off the watch and strapped it on Anna's wrist. Her face beamed and Anna's mother cried. Michelle took several pictures of Anna holding Mickey up next to her face, and then dug into her backpack for several packages of quartz batteries that would be needed to keep the watch chirping. She showed Anna's mother how to open the back of the watch and drop the batteries in, "when he won't talk any more." She wasn't sure she had communicated the message correctly, but Anna said she understood—even if her mother didn't.

" 'Oh, boy!' Mickey" had a home at last. When the bus pulled away from the Hematological Center, Anna's round face was pressed against a second-floor window. She waved goodbye, with Mickey bobbing up and down, firmly strapped to her wrist.

14

PROJECT FRESH AIR

WHILE MARY BETH Ormiston was in Minsk with the Citihope team, she had the opportunity to shake hands and share hugs with the parents of ten children with cancer who later that summer would be coming to Camp Pendalouan, the YMCA camp she directed in western Michigan.

"Those parents are amazing," she told Michelle Carter after they returned to the States. "They seem to have blind faith in a woman they have only just met, in a country they really know very little about. In some cases, they are sending their terminally ill children away for two weeks. I'm not sure I could have done that, but these parents have.

"That's why I feel it's so important that what we do is for the parents as much as for the children. We put together a packet of maps, pictures, and complete information about the camp so they can be as comfortable as possible putting their kids on that plane."

Mary Beth was close to forty, with dark, curly hair and a wide, easy smile. Her earlier camp experience with children of Chernobyl had been so positive that she had decided to travel to Byelorussia with the Citihope group in June.

The previous summer, a group of kids had attended Camp Catch-a-Rainbow, a special session for children with cancer. This year, Mary Beth intended to "mainstream" them into a camp with healthy Michigan kids, in an effort to make their illnesses less of an issue.

"We found that the Chernobyl kids knew very little about their disease," Mary Beth said. "Most didn't even know they had cancer, since doctors in the Soviet Union generally don't tell even their parents the children's prognoses. These kids weren't as comfortable dealing with the disease themselves or dealing with other children who knew about and had confronted their cancer."

Mary Beth's kids had flown into New York just a few hours ahead of her on her return from Minsk. She met them in Manhattan and caught a flight to Michigan with them and their chaperones, Dr. Lydia Litinovich of the Byelorussian Children's Fund and Dr. Reiman Ismail-Zade of the Children's Oncological Institute.

Once they arrived in the Midwest, Mary Beth introduced the children to their host families (with whom they would stay for a few days before and after camp) at Hackley Hospital in Muskegon. There the kids were given brief medical checkups and snacks.

"The medical exams are essentially to verify that the kids don't have any kind of infections or colds or whatever," she explained to the two Byelorussian doctors, "not to go over their cancer treatment programs. We didn't bring them here to interfere with or enlarge upon the medical treatment you are giving them in the Soviet Union."

This year, Mary Beth knew that cookies and Big Macs weren't what the children wanted; she had bowls of fresh fruit—especially bananas—in quantity and, just like the summer before, the kids couldn't get enough.

The third night of camp, long after campers were supposed to be in for the night, Mary Beth and camp director Jerry Courtney were talking over coffee in the dining hall. The door opened quietly and there stood tall, gangly, and shy Igor, one of the older boys. Instinctively, Jerry nodded to him, went to the kitchen, brought out about two dozen bananas, and gave them to the boy. Igor smiled broadly, offered a polite "thank you" in English, and took off running.

"I'm afraid to ask if all those bananas are for him or if he's going to share them with his whole cabin," Mary Beth said with a laugh when Jerry came back to the table.

"It's better that you don't know," he said as he finished his coffee.

Experience with the previous year's camp and her trip to Byelorussia led Mary Beth to make a few other changes. "This year we are taking special pains not to throw our capitalism in their faces the way we did before. We are avoiding trips to the supermarkets and toy stores, and focusing on picnics, the park, the zoo, and the beach. They absolutely love these trips and they're much more beneficial."

Mary Beth was equally careful that Dr. Lydia and Dr. Reiman not feel like they were on display. "They have been able to participate in surgery, go to seminars, and make contacts with a relief agency that's gathering supplies for them. They're making lists of the kinds of equipment and medicines they really need."

Efforts were also made to remember the families left in Byelorussia. "Last year we had a little girl who would stash away anything she was given, candy or a small toy, even clothes. It took us a couple of days to figure out that she was saving everything to take back to her little brother. This year we are putting together 'care' packages for the family—for mom, dad, brothers, and sisters—so the kids can go back and give them each remembrances of the exchange program."

After the camp was over, while she was working on a reunion of all the host families in the fall, Mary Beth found herself humming the refrain from a folk song the Byelorussian children had taught her:

> Let there always be sunshine!
> Let there always be blue sky!
> Let there always be mommy!
> Let there always be me!

❊ ❊ ❊

ABOUT THE SAME TIME, in Sonoma County, about sixty miles north of San Francisco, individual families were preparing to open their homes and hearts to another group of children of Chernobyl.

Late one evening, Cliff and Connie McClain got a call from Michael Christensen. The ten children from Vetka in the contaminated Gomelskaya region they were expecting "sometime in the middle of June" had arrived. They were waiting at the airport.

Michael had received a call from Dr. Olga Volmyanskaya of the Minsk Institute of Radiation Medicine, who had brought the children halfway around the world. They were all at San Francisco International Airport, and no one was there to meet them.

Jolted—but not terribly surprised by the lack of advance warning—Michael shifted into action. He borrowed a van from church, picked up ten tired and pale youngsters, ages ten to twelve, and brought them back to his house in San Francisco. His wife Rebecca cooked them chili and hot dogs, but there were few takers. One-year-old Rachel and Brewer, the dog, delighted the Byelorussian kids until they fell asleep on the living room floor.

Cliff and Connie McClain were ready, though, and they alerted the host families. Soon the weepy and travel-weary youngsters were settled in for six weeks of fresh air and California sun.

ORIGINALLY THE MCCLAINS had been working with a Ukrainian teacher whom they had met in Kiev. She was looking for American homes for Ukrainian children, and the McClains—who were grandparents and also had experience with foreign exchange students—were ready to provide them. However, arrangements had bogged down, and they were beginning to despair that they would ever see Soviet children.

"That's when a neighbor showed me an article in the *San Francisco Examiner* about Citihope taking Thanksgiving turkeys to Minsk," Connie said. "The article mentioned Michael Christensen. So I called him at his church and asked if we could meet. We did, and he promised to pass along to Citihope's partners in Minsk our request to host children-at-risk, not children in treatment.

"When Michael returned after his Christmas trip, he said the Byelorussian officials were only interested in sending children in treatment to the West. So we put everything on hold again."

It wasn't until April 1991, the fifth anniversary of the catastrophe, that Michael called with the name of a man from the Byelorussian Charitable Fund who was willing to send children-at-risk to the United States for a respite. Michael put Connie in touch with him, and the connection for the mid-June arrival was finally made.

"We will be getting ten children, about ten to fourteen years old, for six weeks from mid-June to the end of July," Connie reported to Michael. "We've got the names of the children and their ages but no arrival dates. Let me know if you hear anything."

They found families in three Sonoma County towns—Petaluma, Rohnert Park, and Santa Rosa. They were the kinds of communities where children ride bikes, play baseball, swim, picnic, and play with pets, activities that were severely restricted in their home town of Vetka, about sixty miles from Chernobyl.

As part of Citihope's Project Fresh Air, Michael served as international facilitator for the McClains in Petaluma and the Byelorussian Charitable Fund—which had borrowed the hard currency to pay the children's airfares to San Francisco. The McClains also took on the task of raising the funds to offset those travel costs; by the end of July, they had four thousand dollars in a local bank account.

Six weeks provided plenty of time for the youngsters to make themselves at home with their new American families. They craved fruit of all kinds, especially bananas, apples, and grapes. It took a while for them to learn that they didn't have to stuff themselves, that there was more at the supermarket when the fruit bowl was empty.

They developed a passion for Saturday morning cartoons, computer games, and sandlot baseball. They also got their teeth filled—as many as nine fillings for a single child. A local dentist donated several hundred dollars of dental work for the youngsters, who were coming from a country where preventive dental care was not routinely practiced.

The McClains had arranged for daily English lessons at a local church, but language proved to be an overrated barrier. Although Dr. Volmyanskaya was available by telephone if they needed something important translated, everyone got along very nicely with pointing, gestures, and that international language—kidspeak.

One family had the advantage of language fluency. Liz Pacini, the granddaughter of a Russian émigré who fled the country during the Bolshevik Revolution, and her husband, Richard, took Volya Klimov into their home. Their son, Joey, transferred his own passion for baseball to Volya, who had soon learned the names of all the Major League teams and many of the players. He rarely missed a game on television, and the Pacinis packed two baseball gloves into Volya's bag when he left—one for him and one for a friend so they could play a proper game of "catch."

The youngsters looked like very different kids when they boarded a jet at San Francisco International Airport at the end of July. Were they any healthier? They certainly appeared to be, but no one knew how much their immune systems had been bolstered.

The children weren't even on the plane before the McClains began working on an expanded R & R program for the following year. Connie had also enlisted the help of a group of doctors to see how the people of Petaluma could help provide medical expertise, pharmaceuticals, and medical supplies for the town of Vetka as well.

However, before the summer was over, the McClains received a letter from their earlier contact in the Ukraine, who was ready to send fifteen children in the winter with her own daughter, a pediatrician, as an escort. So the McClains began interviewing host families for a group to come in February.

Even the Petaluma school district would be involved in this project, since the children would be coming during the school year. Connie said the superintendent was excited at the prospect and that the children would be included in regular classes during their stay.

Within twenty-four hours of the newspaper announcement of the impending trip, Connie had twelve telephone calls from families wishing to serve as hosts. "This confirms my belief that people want to do something, but it needs to be something possible," Connie said. "They may not be able to travel around the world with food, but they can take these children into their homes. This is definitely in the realm of the 'possible.' "

❊ ❊ ❊

THROUGH PROJECT FRESH AIR, another group of children—from Narovlia—were getting a dose of Americana in upstate New York. From early July to the end of August, seven youngsters ages ten to fourteen played volleyball on the lawn of the Moore family's home in the Catskill Mountains of New York. They also swam, hiked, and shared television-free evenings with families at the Woodcrest Bruderhof community in Rifton, where they found the dense thickets of trees remarkably similar to the forests that cover two-thirds of Byelorussia.

Paul and Sharon Moore were in the process of moving their lives and radio ministry from Times Square in New York City to the small community of Andes in the Catskills. Paul had resigned his position as associate pastor of St. George's Episcopal Church in Manhattan to devote himself full-time to Citihope and its international relief project for the children of Chernobyl. But for two weeks, the bustle and confusion of the move were put on hold while Paul and Sharon hosted the seven from Narovlia and their interpreter, Irina Gritsenko from Minsk.

The Moores were last-minute substitutes for the families that an Episcopal parish had recruited to host the children for the summer. The ever-frustrating logistical mixups in Minsk had prevented Citihope from giving the church families a definite date of arrival, causing the well-organized Episcopalians to pull out of the project.

Michael Christensen had been ready to send a fax to the Byelorussian Children's Fund, telling them that the children couldn't come after all, when Paul held him back. "Let them come. We'll take them for part of the time, and we'll find other families for the rest of the summer."

The children arrived expecting summer-long stays with individual families. Instead, the Moores were able to deliver only two weeks at their home in Andes, two weeks with another set of families, two weeks at camp, and two weeks at the Hutterian Bruderhof.

"This was a real learning experience," Paul later said. "Not only did the kids get something different from what they expected, but Sharon and I did too. Now we realize that our expectations were unrealistic.

"We had arranged civic programs and receptions, but the kids weren't interested, and for the most part were uncooperative. They only wanted to swim and play—which we should have known." The Moores became frustrated when the kids balked at the role that had been thrust upon them, and the children resisted being shuttled around to four different locations, including a church camp with its unfamiliar rituals.

The lesson was well-learned: contingency planning and the willingness of American hosts to be flexible would be critical to future success.

THE EXPERIENCE at the Bruderhof community saved the

trip for the seven children from Narovlia. Two of them—Andrei, age fourteen, and Lena, age fifteen—spent the entire summer at the Woodcrest Bruderhof, and the rest of the children joined them for the last two weeks of the stay. There they all took pleasure in the simple spirituality that the Hutterians, such as Asa Tribble, practiced daily.

One of the Bruderhof families, Tony and Jenny Potts and their seven children, was chosen by consensus to serve as hosts for Andrei and Lena for the summer. They accepted the will of their community in selecting them, but Tony said, "At first, we wondered how it would go. We soon found out there was nothing to worry about. We were really on the receiving end."

The opportunities for Andrei and Lena were light-years away from the trips to amusement parks, K-marts, and fast-food restaurants that had often punctuated the agendas of other visiting children of Chernobyl. A camping experience in the Adirondacks, free of fears about radiation, was one of the highlights.

The Potts children reported, "Andrei would go out on the lake in the canoe by himself in the early morning, and he would chop wood for the fire. And when we climbed Snowy Mountain, he went all the way up barefoot. He said, 'In the city, I wear shoes, but here on the mountain I go barefoot.'"

❊ ❊ ❊

THE SUMMER WAS EBBING before the last Project Fresh Air group, six children in remission from leukemia, Hodgkin's Disease, and thyroid cancer, arrived at Camp Sunshine Dreams in the Sierra Nevada Mountains of Central California.

There, at Huntington Lake, a ninety-minute uphill drive from Fresno, the Byelorussian campers joined eighty American youngsters for a week as "just kids," under the auspices of the American Cancer Society and Valley Children's Hospital in Fresno.

They had hardly settled in before static-charged radio reports announced that Mikhail Gorbachev had been unseated in a coup in the Soviet Union. But by midweek, the coup had failed. By Friday, the youngsters and their chaperones (Svetlana Lukashuka, whose husband was involved in demonstrations against the coup in Minsk, and Dr. Larissa Danilova) were speculating on how different the country they were returning to would be from the one they had left just two weeks before.

On the last day of camp, Michael Christensen, Michelle Carter, and Faith Ferdinand (who worked with Michael on a number of social justice projects in the Bay Area) left San Francisco at dawn and drove more than five hours to Camp Sunshine Dreams.

The Byelorussian youngsters, three girls and three boys ages seven to sixteen, were indistinguishable from the other campers—except when they were on the pier or in boats on the lake. There they had to wear bright orange life preservers, since none of them could swim. They got dirty, ate tacos and chips, learned to play softball, paddled canoes, and described nearly every new experience as "awesome"—all without anyone thinking twice about missing limbs or bald heads.

Dr. Danilova told Michael that two of the Soviet children had fallen into the cold, snow-fed water of Huntington Lake a couple of days earlier, and had paddled around in their life jackets before they allowed themselves to be rescued. "I can't believe it, but none of them got sick from the cold, cold water," she said. "I was sure they would develop pneumonia."

Sveta Lukashuka also commented, a little wistfully, on the artificial limbs that the California kids were wearing, some of them right into the water. "Here when children have amputations, they get prostheses and they can play; they can come to camp," she told the other counselors. "But in our country children are poorer. It is impossible to get a prosthesis. The children have to stay at home."

At Camp Sunshine Dreams, the campers were encouraged to "take responsibility for their treatment and control of their illness," according to camp director Sunny Shervem. The ceremonial campfire on the last night of camp is part of that process of "taking responsibility."

The final campfire of any camp can get pretty emotional as kids prepare to say their goodbyes until next year. But the "spirit stick" ceremony at Camp Sunshine Dreams was enough to break your heart.

After the familiar array of corny skits and silly campfire songs, the giggles and taunts subsided as the camp director invited campers to come forward and place a special "spirit stick" on the campfire and sprinkle shavings from that stick over the flames as they shared their personal camp experiences. Inevitably, they talked about having cancer, losing siblings, growing up (or not growing up), and the campers who were no longer with them.

The ceremony was a chance for campers to express their feelings and confront the reality of their condition. For the Soviet campers, it may well have been the first time that they had been able to know the truth about their disease and deal with it. Now they knew. And so did their siblings, who comprised about half of the campers. "It's important for the siblings to be included," Sunny explained. "They are victims as much as their brothers and sisters, and they have a lot of issues to work out — the usual sibling rivalries, which are all mixed up with envy and guilt." Siblings were invited to camp even if their brothers or sisters had died.

The silence around the campfire was thick as one small, shaky voice after another spoke up. This year one of the voices was Russian. Sixteen-year-old Larissa recalled other children of Chernobyl at hospitals in Minsk, and Sveta translated. Larissa knew some English, but she was too emotional to try using it.

Larissa had been the one to seek out Paul Moore when he met the group at JFK Airport and took them on a whirlwind sightseeing tour of New York before they left for Los Angeles. She had obviously been paying attention, months earlier, when Dr. Olga Aleinikova had told her that the methotrexate she was getting as treatment for her leukemia had been a gift from the "tall priest from America." When Larissa saw Paul, she told him that she was well because of the medicine he had brought to the Children's Hematological Hospital.

The spirit stick ceremony ended in sobs and hugs and the soft verses of "Kum ba yah." The Byelorussian kids, Sveta, and Dr. Danilova held onto each other in one big, disorderly hug for awhile, and then opened the circle to embrace their new friends.

Michelle and young Larissa had gotten to know each other during that day, thanks to Michelle's sometimes chancy Russian and Larissa's bit of English. They embraced each other and, through her tears, Michelle brushed back Larissa's stylishly cut blonde hair, saying, *"Lara, bog lyubeet tebya ee muee lyubeem tebya"* (God loves you and we love you).

As the tears gave way to roasted marshmallows and "s'mores," Michael's friend Faith noticed that the youngest of the

Soviet campers hadn't joined the others. She went off in search of seven-year-old Andrei Aleksandrovich, who had had some rough moments the past week. This was the same youngster who had bravely submitted to a second withdrawal of bone marrow so he could provide Sharon Moore with a fresh set of slides. (As it turned out, oncologists in the United States reaffirmed the diagnosis of his own doctors in Minsk.) Andrei was very young to be so far away from home and so sick.

Faith found him lying on his bunk with his face to the wall. Despite the fact that she couldn't speak Russian, she used the universal language of hugs and touch to entice him back to the campfire. Then she showed him how to put marshmallows on the end of a stick and burn them to a charcoal mass before popping them into his mouth.

Despite some serious bouts of homesickness, Andrei did manage to have some fun that week. Small for his age, with fine features and short, sandy hair, he was noticed by a reporter for the *San Francisco Examiner*. She wrote about how Andrei "stepped up to bat, swung mightily and smacked the softball. . . . Never mind that the little guy scampered toward the pitcher instead of first base. And never mind that he was tagged out when he was eventually rerouted to first. The team instantly dubbed Andrei 'Giant Slugger' and the umpire gave him a high-five. . . .

"When the teams changed positions, Andrei stayed in the batter's box. This being a rather informal game, he was made designated hitter for both sides. . . . As the innings passed, Andrei drifted away from the game, evidently convinced that the American pastime is a trifle tedious."

PART FIVE

*The three-day putsch that briefly deposed Mikhail Gorbachev pro-
vided a world stage for the democrats to dramatize their power and Boris
Yeltsin to claim the spotlight. In one republic after another, democrats
found the power to shake off communist shackles and assert their
strength—and in some cases their independence—in a fury of national-
ism and ethnic expression.*

*It soon became clear that the celebrating democrats were dancing on
the grave of the Soviet Union. The autocratic, seventy-three-year-old
workers' paradise, built on the premise of a constant struggle toward a
perfect communist state, was crumbling into far more than fifteen disparate pieces. Gorbachev, the leader who cautiously tipped those first domi-
noes and then watched them race far beyond his grasp, lost his country
and his job.*

On the eve of novee got, *the union ended with the old year—and so
did the grand charade. The world peered into the Soviet coffin and saw
an emaciated old man with withered limbs and missing teeth.*

*And what of the old man's offspring? Lithuania, Latvia, and Esto-
nia turned their backs; they had never been legitimate Soviet children,
and their "adoption" had been illegal and immoral. But would the Russian bear gather the rest of its former siblings under her protection? Did
Ukraine, Georgia, Armenia, and Kazakhstan want to be protected? Did
Belarus?*

*They did not. They all remembered too well the arms of the bear
around them—an embrace that had felt more like a stranglehold. They
would each find their own way—together but separate, at arm's length
from the bear.*

15

"WE'VE PAID A TERRIBLE PRICE"

IT WAS OCTOBER and, in the emerging nation of Belarus, the euphoria of political change was beginning to take on the hard edge of reality. Prices were creeping up, and demands for hard currency instead of the shaky ruble were heard everywhere. The same people who had waved the white-red-white Popular Front banner in September to cheer the demise of the Communist party were now grousing about the price of their newfound liberty. The police state had at least been predictable; the uncertainty of their future was now both exhilarating and oppressive.

Out of this personal and political confusion came a fax from Dr. Olga Aleinikova: "Please consider this most desperate appeal. Everything here is in crisis. Our hospital needs the most basic items as well as methotrexate, oncovorin, leukovorin, and vincristine. We must have antibiotics, penicillin, and gamma globulin. Right now we are able to buy almost nothing."

Dr. Olga faxed that frantic plea to Paul Maxey, the pharmaceuticals broker at Allied Medical Ministries in Maryland. For nearly a year, he had been maintaining a "wish list" for her hospital in Minsk.

This time Dr. Olga needed the basics, the drugs she had always been able to obtain one way or another before. But these were new and harder times, and even the basics were impossible to get.

What remained of the Soviet Union was a jumble of neo-nations with no functioning economic system, no exchangeable currency, a centralized plan for industry and public services that had no center, and a health delivery system that hadn't worked in decades.

In the midst of this chaos, Belarus still had 2.2 million people, including 800,000 children, at risk from the legacy of Chernobyl. Conditions that had been difficult a year earlier, when Pyotr Kravchanka quoted the Book of Revelation to the United Nations General Assembly, were now critical. Available resources, which once were thin, no longer existed.

Eastern Europe had previously provided the majority of pharmaceuticals and medical supplies for the Soviet Union. Payment was now demanded in hard currency instead of tractor parts and television sets. The remaining supplies that had come from other republics within the union were equally impossible to obtain.

❅ ❅ ❅

ONCE AGAIN, the Paul-to-Paul connection kicked in. While Paul Maxey worked the phones to find the necessary drugs, Paul Moore raised the funds and arranged transport—and considered the future.

"Michael, we both know the Soviet Union will no longer exist soon. It's just a matter of time. I want us to be there when it finally collapses, to help redeem the situation." Paul was calling from his still-unpacked office in Andes in upstate New York.

"We also know that the intimate, person-to-person projects like the ones we've been running for the past year will never meet this new, escalating level of need. We have to start thinking in bigger terms, much bigger."

Michael listened from his small home office in the back of the Laird-Christensens' narrow Victorian house in San Francisco. "How big a project do you have in mind, Paul?" Michael was wary. He knew how Paul's ideas could be spectacularly successful when they hit the mark, and bureaucratic messes when they didn't. If it didn't work, Michael knew he would be asked to help clean up while Paul started dreaming up another project.

"The Bush Administration is ready to buy into the program," Paul said. "Massive food and medical relief efforts like Project Hope and AmeriCares are suddenly infused with federal money and high visibility, and we are right in the middle of the effort. I was in Washington earlier this week to talk with the Bush people and the leadership of all these relief programs. In nearly every case, we've been able to get Belarus added to the list of distribution sites.

"The time is right," Paul said. "The media are taking care of that for us, now that the crisis in the Soviet Union is front-page news and revelations about the unraveling of the social fabric are breaking every day."

Paul outlined a plan for a massive air shipment. He had already obtained a warehouse, and pharmaceutical companies were filling it with sizable donations.

"Aeroflot has promised a plane to pick up as much as one hundred and fifty tons of cargo—the fax just came this morning—but I have to get it palletized and ready to go. I'm going to have to hire a warehouse crew to manage all that."

"Sounds like Citihope has moved into the big leagues now," Michael said. "Keep in touch and let me know if I can help from here."

Michael knew that negotiations with the Russians never went smoothly, but even he was surprised when the next call came. Paul sounded frustrated, exhausted, and about as low as he had ever heard him.

"I've had a hundred and fifty tons of medical supplies and pharmaceuticals sitting in that unheated warehouse since October 21, and there's no plane. I have all the faxes from Aeroflot with all the appropriate promises, but we still don't have a plane or even an idea when we might get one.

"At this point, I'm convinced that Aeroflot is trying to sabotage the flight. It's some kind of a control issue among different groups of bureaucrats—probably all well-meaning. With all the republics declaring independence, no one's sure who even owns Aeroflot.

"Whatever the reason, Aeroflot won't release funds for refueling the plane en route. They want us to pay eighty thousand dollars in hard currency up front for the fuel."

In the climate of confusion and rapid change, even the Soviet national airline wanted to deal only in hard currency.

An entire month passed with thirteen truckloads of sensitive medical cargo sitting in a warehouse while the Moores waited for the Soviets to claim their relief shipment.

"I've pulled every string I can think of. I talked to Colin Powell. Senator D'Amato even gave a speech on the floor of the Senate urging the Air Force to provide a cargo jet for the relief flight. The national media have been up here on our doorstep for the past week."

Help finally arrived from a familiar source. Nadia Matiwsky of the Ukrainian-based Chernobyl Children's Relief Fund had been working with Aeroflot to get an Antonov-225, the largest plane in the world, to the United States as part of a fund-raising strategy. People would tour the plane and the fund would raise money for the children of Chernobyl.

Just four days before Thanksgiving, she was able to arrange for the huge jet to carry Citihope's relief cargo to Minsk.

❋ ❋ ❋

"WE HAD ONLY a couple of days to get it all trucked to the Newark airport and meet the plane. Of course, it all had to be palletized and ready to be loaded," Paul told his radio audience. "We worked around the clock, and at 3:00 A.M, before we took off at 6:00 A.M., I managed to break my foot. But, bandaged and on crutches, I still got on board."

So, just one year after Sharon and three other Citihope volunteers had brought dozens of frozen turkeys to celebrate an American-style Thanksgiving at an orphanage in Minsk, the Moore family was aloft in the world's largest plane, carrying one hundred and fifty tons of medicine and medical supplies to the children of Chernobyl.

Paul, Sharon, and PJ were seated on benches along the bulkheads of this huge plane, which clearly hadn't been designed for passenger comfort. They were making the best of the long flight when the plane began to lose altitude. The crew, with noticeable beads of sweat on their faces, began frantically shifting cargo and shouting orders to each other.

"What's going on here?" PJ asked his dad. "They act like they're scared, like they think we're going to crash."

"Paul, see if someone in Nadia's group up front knows what's going on," Sharon said. "Something is definitely wrong and it's terrible not being able to understand what they're saying."

Just then an English-speaking member of the Chernobyl Children's Relief Fund made her way back to the Moores. "The crew just told us that we're going to have to make an emergency landing in Prague. The plane's hydraulic system isn't working properly."

Paul had always imagined himself behaving heroically in a moment of crisis, gathering all the passengers and crew together to pray and delivering the sermon of his life as the plane descended through the clouds.

But now that he and his family were in mortal danger, he wanted only to gather them around himself and hold on. Paul, Sharon, and PJ held hands and found themselves singing familiar hymns through clenched jaws. Meanwhile the CCRF delegates in the front and the crew members strapped themselves into safety restraints along the bulkhead.

No calm, reassuring flight attendant appeared to suggest that they remove high-heeled shoes and glasses, put their heads down on their knees, and cover their necks with their hands. But the three Americans had seen enough Hollywood versions of airplane crashes to react instinctively.

As the shuddering Antonov approached a foam-covered runway lined with emergency vehicles just outside the capital of Czechoslovakia, everyone in the plane was absolutely silent. Paul didn't make any special deals with God, but the sense of relief was palpable when the plane landed hard, bounced twice, and slid to a stop, still upright and in one piece.

Shaken but safe, they applauded the pilot's skill. "Hey, I've had worse landings on a normal flight!" PJ cheered.

Unbelievably, the next morning all three climbed aboard the same plane and set out again. They were told that the hydraulic system had been repaired, but now they were on their way to Kiev, where the Antonov was stationed, rather than Minsk—a change in destination that was never fully explained.

After a dozen more crises in sub-zero temperatures that threatened to freeze the medicine, the trip ended with a long, nighttime journey to Gomel in a cramped car. In the end, the Moores saw their cargo delivered to eleven different hospitals in the Chernobyl region of Belarus, and hosted a second annual Thanksgiving dinner (with twenty-four frozen turkeys) at a sanatorium in Gomel.

✷ ✷ ✷

JUST BEFORE they were scheduled to fly back to New York,

the Moores were summoned to Minsk for a meeting with the new president of the republic, Stanislav Shushkevich.

"We found ourselves in the same cabinet room where former President Dementei had presented the medal to us a year earlier," Paul spoke into his portable cassette recorder later that night. "The huge portrait of Lenin staring down at us was gone, and in its place was a portrait of Francisco Scarina, the Belarusan patriot and printer who translated the Bible into Byelorussian."

Paul and Sharon hadn't a clue as to why Shushkevich wanted to see them. "He looked like Winston Churchill—round, ruddy, warm, but not fuzzy. He was an elder statesman and he exuded sincerity."

After some flattering chit-chat about Citihope's good deeds, Shushkevich got down to business. "I know you've come to help us, and I feel you should know that there are three major problems that we face at this critical point in our history.

"The most important way you can help us is with our first problem. We need to know God. We've been raised not to believe in God, to believe only in the state. Therefore our children have no spiritual roots. They don't know right from wrong. They have no moral guidance, and we have paid a terrible price for this. You must show us the way to know God and help bring spiritual life to our country.

"The second problem we face is the need for food. But we are a resourceful people and I believe we can solve this ourselves.

"The third problem is our need for medicine and pharmaceuticals, since we don't make them here and we can't afford to buy them. We desperately need your help on the first and the third problems.

"As for me, more than anything else, I need your prayers."

Paul led them in prayer and Sharon shared a psalm of encouragement. Finally, Paul felt secure enough to broach the constant logistics problems Citihope had encountered in bringing supplies into Minsk.

"There is no excuse for this nightmare," Shushkevich replied without hesitation. "I will appoint a deputy minister and a committee to work with you to solve all these problems."

Later, at dinner, one of Shushkevich's deputies produced a "memo of understanding" among Citihope International, the Byelorussian Children's Fund, and the Belarusan government. The document promised that the government would "receive, unload, and protect the humanitarian cargo and set up a storage facility that Citihope will monitor. Furthermore, Citihope will oversee the distribution of the cargo, with the Children's Fund serving as a partner with the Ministry of Health in this endeavor."

Vladimir Lipsky and Pyotr Kravchanka joined the Moores for dinner to celebrate and solidify the humanitarian partnership of the Byelorussian Children's Fund, Citihope International, and the new Belarusan government.

Then, with ceremonial flourishes, they all put pen to paper to sign the new contract. Lipsky, who had watched the development of this partnership over the past year, grinned and patted everyone on the back.

Paul had expected Shushkevich to be at the dinner as well. He was told that he had a high-level meeting to attend in Brest. Several days passed before Paul realized that this session in Brest was the historic meeting of the leaders of eleven of the former republics, which wrote the script for the dissolution of the Soviet Union and the creation of the Commonwealth of Independent States. With a cautious eye on the Russian bear, the new nations chose Minsk (whose chief attribute was that it was not Moscow) as the capital of the CIS.

By coincidence, the birth of the CIS and the establishment of a Citihope warehouse in Minsk had both taken place on the Moores' twenty-ninth wedding anniversary.

16

NOTHING CHANGES, EVERYTHING CHANGES

AS CHRISTMAS approached, and with it the season of sharing, the Western world was focused on the harsh Russian winter and the needs of families struggling in the wake of the disintegration of the Soviet Union. The Bush administration responded to assurances of a peaceful transition to democracy and the continued reduction of nuclear arsenals with support and military transport for private and public humanitarian relief efforts. Massive airlifts bound for Moscow and St. Petersburg were becoming commonplace.

Citihope, which had just joined the big league of relief efforts, remained focused on the needs of the children of Chernobyl and Belarus. It also continued to find value in small delegations that brought people who cared face-to-face with people in need.

Michael had been working for months to recruit a select group of volunteers eager to spend the New Year and Orthodox Christmas holiday in the newly independent republic. He wanted to continue the smaller citizen diplomacy and relief efforts that had put Citihope on the map.

On New Year's Eve, fifteen American volunteers were on the train from Moscow to Minsk. It was snowing early on the morning of December 31—the last official day of the Union of Soviet Socialist Republics—when the Americans arrived in the capital of the Commonwealth of Independent States.

Lenin Square had become Independence Square after the general strikes and popular uprisings of the previous year. Gone was the hammer-and-sickle flag over the Parliament House, and in its place waved the fresh and hopeful white-red-white banner of twelfth-century Belarus, the ancient land of the Slavs.

Food and medicine were still in short supply, and citizens continued to line up to buy whatever goods and services were available. But that night would be *novee got* and the time for celebrating all things new.

The year before, the Citihope delegation had spent the New Year's holiday with Communist leaders behind iron gates in a *dacha* reserved for party *apparatchik* and their guests. This year, volunteers would stay with democrats in private homes and share the holiday festivities with their host families.

Michael and Michelle, coleaders of the delegation, would stay with friends from a previous visit—Svetlana and Alexander Lukashuk, two young professionals who had been active in the Popular Front, their twelve-year-old daughter Katya, and their nine-year-old son Misha.

Alexander, known as Sasha, was a journalist and book publisher who had helped break the story of Stalin's mass executions at Kurapaty outside Minsk. He had also published Lee Harvey Oswald's diary of his days in Minsk in the early sixties as a guest of the KGB. Svetlana was a teacher who had served as interpreter for the children at Camp Sunshine Dreams in California. It was a reunion of friends when the Californians arrived.

Svetlana had planned a traditional holiday dinner and celebration for midnight, complete with a New Year's tree, gift-giving, and a surprise visit. But first the weary travelers needed a late afternoon nap.

Michelle had fallen asleep as soon as she pulled up the down comforter on the bed that Katya had surrendered for the duration of the visit. She was dreaming deeply when she began hearing bells. It took a while to break through, but finally she realized the bells were in the apartment, not in her dream.

She got up and opened the door to Katya's room a bit. There in the hall was Grandfather Frost in his long, red, velvet coat and white beard, with the Snow Princess Snegochka at his side. They rang bells and sang a Belarusan carol.

Michelle couldn't let Michael sleep through this, so she slipped out, knocked on his door, and woke him up. They both came out to join the party just as Grandfather Frost was passing out small toys to Misha and Katya.

Later, during dinner, the television droned on. Michael was interested in seeing how the three stations covered the emergence of the new commonwealth. On the old Soviet channel, now serving the interests of the CIS, six popular comedians joked about recent changes and what might lie ahead. They impersonated People's Deputies who were ordered to study the issues facing the commonwealth and report back on April 1 — the day of practical jokes.

On the Russian channel, a frenzied rock and roll band saluted the New Year. On the Belarusan channel, a growing nationalistic movement was nourished by folk music ensembles in traditional ethnic costumes.

At ten minutes before midnight, Belarusan president Stanislav Shushkevich offered his first New Year's speech:

> I'm very sad to have to tell you bad news on such a happy occasion. Tomorrow we will have higher prices and the economic situation will be very difficult for us all. . . . In the past, we blamed the communists for all that was wrong. Now, some are beginning to blame the democrats because things have not gotten better. But I must tell you that this present difficulty is necessary as we reorganize society and create a new situation in which private ownership is possible. Only with private ownership and a free market can our people work effectively. So let us be patient and unite so that we can make our life happy ourselves. We have known hardship but we have never known defeat.

The Lukashuks were as anxious as any of their neighbors about the economic uncertainties, but their pride in establishing a democracy outweighed their concern. "If I had to choose between freedom and low prices," Svetlana said, "I would choose freedom without a moment's hesitation. Everything dear has a price, and we are paying ours."

Michael and Michelle spent the days between New Year's and Christmas in the contaminated region with the rest of the Americans. They stopped in Vetka, where delegate Connie McClain met the parents of the children who had spent the previous summer with her and her neighbors in Petaluma.

In Narovlia they delivered antibiotics and medical supplies to Dr. Adam Nikonchuk's district hospital. As they carried in boxes of penicillin, the head nurse picked one up and hurried off to the medical wards. As it turned out, the hospital had used its last vial of penicillin the evening before.

In Gomel, Michael worked out the details of the World Vision program that would use the principles of mental and spiritual health to combat the debilitating effects of radiophobia and daily life in the contaminated zone.

❊ ❊ ❊

LATE ON JANUARY 6, as midnight approached, Michael, Michelle, the Lukashuks, and their guests were seated around the dinner table in the Lukashuks' apartment. The discussion centered on the dramatic changes in the country. Michelle quoted a woman she had met on the bus that day: " 'I don't care who's in charge,' she said, 'even the devil himself, as long as I can buy sausage and bread.' "

"I hate to hear that," Sveta answered. "I would rather stand in line and pay ten times more for bread than return to the darkness of the past."

Then Vladimir Lipsky turned to Michael and said, "It's Christmas soon. Will you not offer the bread and the wine?" Lipsky, who had taken his first communion a year earlier when Paul, Michael, and PJ served him the elements at their Christmas party, recalled the impact of that event on his life. He had considered himself a Christian from that day forward.

Michael had not been planning to lead a service that night; he was looking ahead to worshiping on Christmas Day at a three- hundred-year-old cathedral outside Minsk. Nothing was prepared, but everyone, especially young Katya, joined in Lipsky's request.

So the coffee table in the living room became an altar, fresh, black Russian bread was brought from the kitchen, and dinner wine was poured into a red, black, and gold lacquered chalice. Michael found a silk scarf to serve as a stole, and blessed the elements as the group gathered in a circle around the makeshift altar. He put the stole around his neck and explained to Katya, who had just begun attending an Evangelical Christian storefront church in Minsk, that it represented the yoke of Jesus, the task of being a disciple. He quoted Matthew 11:28-29:

> Come to me, all you that are weary and are carrying heavy burdens,
> and I will give you rest. Take my yoke upon you, and learn from me; for
> I am gentle and humble in heart, and you will find rest for your souls.

Sasha, the skeptic of the family, was assigned the role of reading the Christmas story from the gospel of Luke out of Sveta's Russian Bible, while Katya read it in English from Michelle's Bible. As they stood in a circle in the dimly lit living room, where a small New Year's tree dominated the scene, Michael reminded them, "This small group—or any group gathered in his name—represents the family of Christ."

They passed the bread ("His body broken for you") and the wine ("His blood spilled for you") and served each other the elements of the Lord's Supper.

Then they linked hands and prayed: "Lord, give us the guidance and strength to move safely through the uncertain and unknown future that awaits this family and these friends in this new nation at the dawn of this new year. Amen."

❊ ❊ ❊

THE NEXT MORNING, Michael and Michelle were both awake long before anyone else in the apartment, making coffee in the kitchen.

"Last night, Sveta and I were talking after the children went to bed," Michelle said. "She told me that, despite all the excitement of the holiday, the last thing Katya said before she went to sleep was, 'I'm so tired! What if I sleep too late to get bread in the morning?'

"Sveta was upset. She said, 'What's happening to our country when twelve-year-old girls have to worry about standing in line for the family's bread?' That bothered me, too, but I have an idea. We're both up. Why don't we go buy the bread?"

"Why not?" Michael replied.

They bundled up in their winter coats, scarves, gloves, and authentic Russian fur hats (hers borrowed, his a brand-new purchase), and walked—very carefully, since Californians aren't terribly surefooted on ice-covered sidewalks—the short distance to the *produktee*, which was already filling up with Christmas Day shoppers. They picked out two fresh, round loaves of *khleb*, one black and one white.

"You know what we forgot, don't you?" Michelle said as they stood in line to pay. "We didn't bring a bag. Russians never leave their house without a 'perhaps' bag—perhaps they might need it. Stores don't give them to you."

"Well, I guess we'll just have to carry them home in our hands," Michael said. And so they did, feeling a little foolish and looking very much like Americans who were playing at being Belarusans on Christmas Day.

* * *

AT MID-MORNING, Sasha took Michelle and Michael to join the other delegates at the Church of the Transfiguration in a small town outside Minsk. A marathon Christmas service had begun the night before, and the Americans arrived for the last two hours of worship.

They were ushered through the crowd and led inside the altar rail to a spot in front of the gold iconostasis. From there they could watch as one parishioner after another came forward to make confession. The priest would cover the head of each penitent with his stole, then absolve and bless each one.

Nick, one of the members of the American delegation, was so moved that he went forward to offer confession. The priest motioned for Sasha to join them and interpret. "How long since your last confession?" the priest asked.

"Twenty-two years." Nick confessed that he was a lapsed Catholic. With twenty-two years' worth of sins, it took a while to hear his confession. Sasha was self-conscious and uncomfortable in his role as interpreter, but Nick was serious. He reclaimed his childhood faith on Christmas Day.

Finally, communion was served in traditional Orthodox fashion. The priest dipped a silver spoon into the common cup, gathered a morsel of bread and a sip of wine, and placed the elements in the mouth of each parishioner. Even small children were spoon-fed the host. The Americans were amazed when they were invited to partake. Neither Michelle nor Michael had ever been allowed to take communion in an Orthodox church.

At a tea in the parish house after the service, Michael asked the priest, Father Pavel, about his parish. "Your church is truly a church for the people. Thank you for including us in communion. This is the first time I've ever received communion in an Orthodox church."

Father Pavel offered a wry smile. "I am a young priest and I don't know all the old rules."

"How old is your church?" Michael asked.

"The Church of the Transfiguration was consecrated in 1792 and was never closed, even in the darkest days. My father was the priest when I was a child, but I was still not allowed to attend.

Now we have two thousand in church each week, new baptisms every week, and a Sunday school for all the children."

"This is your bicentennial year," Michael noted. "How will you celebrate?"

"We want to restore the cross on top of the dome. With a large cross on top, people will be able to see it from the highway and know that they are welcome. I already have the copper, but I lack the gold to encase it and the skilled labor to build it. At the old prices, it will cost twenty-five thousand rubles (two hundred and fifty dollars)."

"Perhaps the members of our delegation would like to help you." Michael picked up his fur hat and passed it around the table. Everyone, including the interpreters and some German guests, chipped in. When it came back to Michael, two hundred and fifty dollars was counted out and handed to Father Pavel, who promptly praised God for a Christmas miracle!

❄ ❄ ❄

BEFORE LEAVING BELARUS, Michael needed to spend some of Citihope's store of accumulated good will in a rescue effort for a California couple.

For several months, John and Mary Seregow had been trying to get their grandniece, eight-year-old Inna, out of the contaminated region near Vetka so they could arrange for a medical evaluation of her kidney condition in the United States. Long-distance efforts from California had been fruitless, and they turned to Citihope for help. Michael suggested they join the Christmas delegation and see what they could accomplish in person.

Now that Belarus was no longer part of the Soviet Union, the foreign ministry in Minsk was much more receptive to their request. Their affiliation with Citihope was another point in their favor. The documents were arranged and a passport issued to Inna in a truly remarkable three days.

The Seregows still had to deal with the American embassy in Moscow to get a temporary visa for Inna, but when the Americans left from Sheremetyevo Airport at the end of their journey, a very shy and wide-eyed eight-year-old was taking her first airplane ride—halfway around the world!

17

THE HOPE EXPRESS

AFTER PRESIDENT BUSH announced the massive relief effort that would send food and medical supplies to the former Soviet republics, the U.S. Department of Agriculture responded positively to Paul Moore's $2.4 million proposal to deliver 3,500 metric tons of flour, rice, sugar, cooking oil, powdered milk, dried fruit, and baby formula to six cities in Belarus. Paul also proposed to use "ordinary American volunteers" to distribute food parcels to 175,000 families and 63 institutions.

"This is bigger than anything we have ever tried in our lives," Paul warned Michael as he tried to enlist him to head up volunteer recruitment and management. "I'm scared. I need help."

"How many volunteers do you think you'll need, and how much time do we have to recruit them?" Michael asked.

"Eighty volunteers, two months."

Michael swallowed hard. "I really have to think about this. I'll get back to you in a day or so." Paul agreed to wait for Michael's decision, but he knew he had him. Michael wouldn't allow himself to miss out on an opportunity like that.

The sheer volume of cargo the USDA wanted Citihope to deliver was staggering to calculate, let along contemplate. Thirty-five hundred metric tons was 7.7 million pounds of food (about equal in weight to 3,500 Toyotas)—packed into 230 containers that would fill two 100-car freight trains. Stacked 25 feet high, it would cover four city blocks.

"Now that's a lot of relief," Michael said as he tossed his calculator onto his desk and leaned back in his chair.

To handle the logistics of shipping the USDA provisions to Belarus and to design an efficient, black-market-proof distribution system, Paul turned to his old friend John Mumford, a former Navy officer with an available network of retired military personnel.

"This really is unprecedented," Mumford responded. "I'm not aware of the USDA ever having been willing to use American volunteers for distribution. Usually they ship relief to an area and leave it for local agencies to distribute. This should be a unique challenge."

The result was Operation *Nadezhda* Express (Operation Hope Express), an ambitious citizen-to-citizen diplomacy effort intended to put food directly into the hands of the people who needed it most. Michael and his associate, Veronica Hernandez, were able to recruit sixty-three volunteers—doctors, nurses, teachers, scientists, lawyers, journalists, students, and retirees— who were willing to give two weeks of their lives and pay twelve hundred dollars for the privilege of doing so.

Rebecca Laird-Christensen, five months pregnant with their second child, was one of the volunteers. She described the operation in Minsk: "Daily we were bused to a huge parking lot filled with semi-trucks loaded with food that had been shipped across the Atlantic, railroaded across Europe, and trucked from Poland. Two-by-two, we were paired with the director of a local charitable organization and sent off with a truck to count boxes and offer citizen's diplomacy."

Michelle Carter led a delegation of twelve, including husband Laurie and Mike Venturino, to the most distant site, the border city of Brest. Fifty-one years after the Germans had launched their attack on the Soviet Union from this city, twelve Americans were launching an invasion of another sort.

Michelle took extensive notes that would become part of a newspaper story when she got home:

> *Soon we were standing in cold rain in a muddy village thirty kilometers from the Polish border to unload 882 boxes, each holding forty pounds of food, into vans that would head out to even more remote surrounding hamlets. We sat down to tea in the flat of a family of "Chernobyl resettlers" who had been moved from their family home in the contaminated Chernobyl Crescent to a new, sterile home far away from family, roots—or a job of any kind. We collected bear hugs from berib-*

boned veterans of the Great Patriotic War and teary kisses from babushkee with hearts—and teeth—of gold.

And we faced down—and then won over—a proud and overzealous militiaman who put his hand over camera lenses to keep the volunteers from making a record of his people "begging."

"Our people don't beg," he shouted, but later he came back with smiles and pins to trade.

The volunteers, coordinated by Michael and Veronica from the nerve center of the operation, Room 820 of the Hotel Belarus in Minsk, were dispatched to six different distribution sites. Other volunteers set out for Narovlia, Gomel, Mogilev, and Bragin. Additional delegations would handle distribution in Minsk and the surrounding area.

The logistics of getting two hundred and thirty truck-trailers to orphanages, hospitals, prisons, and individual families had been worked out in advance by Mumford and his skilled group of retired military officers called the Washington Group, under contract from Citihope. Coordinators from the group were waiting at the distribution sites for the volunteers to arrive, and then sent them off into the countryside with military precision.

Military precision, of course, hadn't counted on a power outage at the loading site in Warsaw, or sabotage by Polish truck drivers who were peeved at the trucks with the *Nadezhda* Express logo moving unimpeded through customs at the border. These glitches were particularly galling after the cargo had moved flawlessly from Norfolk, Virginia, on two container ships to Bremerhaven, Germany, and then by train to Warsaw.

"Michael, can you hear me?" Michelle shouted into the red telephone in the lobby of the Hotel Druzhba in Brest, the only telephone in the hotel on which long-distance calls could be made. When she finally got through to Minsk—after dozens of failures—the connection was bad.

She cupped her hand around the mouthpiece and shouted as loudly and distinctly as she could. "We still don't have any trucks. We are going to be late in getting back to Minsk. Do you want us to stay and complete the job, or come back in time to join the other teams?"

She had to repeat everything several times, but when the conversation was over, the decision was made to do double-duty when the trucks did arrive. When it could be managed, volunteers would be sent out twice a day to make their deliveries, some to sites that were six or seven hours of driving time away. There families were waiting—families that local social service organizations had determined met the stiff need requirements: a monthly income below one thousand rubles, and/or they were hospitalized, disabled, war veterans, orphans, victims of Chernobyl, or Chernobyl "resettlers."

When the trucks finally did arrive, time and logistics required most volunteers to go out alone, with only a Polish-speaking truck driver, to meet Russian-speaking strangers in a totally foreign village.

Mike Venturino was the first American ever to set foot in the city of Stolin, and he became a celebrity of such proportions that when Laurie Carter returned two days later with another truck-load, he said, "I felt like Buzz Aldrin, the second man on the moon."

Michelle rode out alone to the village of Zhabinka "in a truck cab with every available surface covered with pictures of nude women," she later told Laurie. "The coup de grace was a set of foam rubber breasts that dangled over the windshield in front of me for the entire trip. The driver saw me staring at the jiggling breasts, and he smiled, using the only English words he knew: 'Nice, huh!'

"I've begun to think of the experience as something of a metaphor for our stay in the beautiful city of Brest."

Members of Michelle's outpost delegation ranged from a twenty-four-year-old Minnesotan to a seventy-year-old French émigré. Over and over, the hosts in the remote villages were moved to see grandmothers and grandfathers among the Americans climbing down (and occasionally tumbling out of) the trucks.

Part of their ritual in each village was an explanation that the volunteers really were "ordinary" Americans who didn't work for the government, who were using their own vacations and paying their own way. To these people who rarely had visitors from Minsk, let alone Minneapolis, that spoke volumes.

Beyond the value of the food itself, the greater impact of the effort was the hope it represented—hope that the villagers translated into support for them during a very difficult transition from totalitarianism to democratic nationhood.

These "ordinary" American volunteers had been recruited to hand out boxes of food to proud, self-reliant people who had lost a quarter of their population during the four-year Nazi occupation, had absorbed seventy percent of the radioactive fallout from Chernobyl, and were trying to feed and clothe their families in a time of triple-digit inflation.

"But what we really brought with us," Michelle later wrote, "was that quirky American commodity—noisy, bumbling, smiling, compassionate people who cared enough to come."

❊ ❊ ❊

WHILE THE VOLUNTEERS were unloading trucks in cities like Baronovichi, Ivanova, and Pinsk, Paul Moore had another agenda back in Minsk. He had his heart set on a helicopter ride to a couple of the distribution sites so that a film crew could record the deliveries. He asked Belarusan government representatives and came up empty; then he turned to Alexander Trukhan of the Byelorussian Children's Fund.

Apparently Trukhan was still well-connected to the old Soviet military and party apparatus. His good friend was commander of a once-secret military base outside Minsk; the chopper was secured, and the Moore family and Michael were joined on board by assorted Belarusan dignitaries, an interpreter, and the film crew.

A car picked them up at the Hotel Belarus and ferried them out of the city, through several guard posts, a tunnel, and onto a military base secluded in an evergreen forest. There, nestled in the woods, was a small airport with transport planes and helicopters on the tarmac.

As they were led up to one of the combat helicopters, Michael asked Alexander Vasiliev, deputy Belarusan ambassador to the United States, "Is it all right to take pictures?"

Alexander shrugged his shoulders. "There's no more Soviet Union and, therefore, no more rules."

As he walked across the base, Michael was sure this would never have been permitted in normal times. It was tolerated here only because of the collapse of the union and the subsequent absence of new policies and lines of authority.

Right now, anything goes, he thought, *rather like the frontier days of the American West.*

Before they climbed in the helicopter, Sharon Moore asked if she could put an Operation *Nadezhda* Express poster on the side of the craft, identifying this as a humanitarian aid mission. Permission was granted—with obvious delight. So the operation logo—a large red, white, and blue locomotive with the American and Belarusan flags and a banner reading " 'USDA,' 'In God We Trust,' " in Russian—took up a conspicuous spot right next to the red Soviet star.

Five one-hundred-pound boxes of medicine, representative of many more boxes coming by truck, were loaded on, and the delegation was off to the city of Gomel. At a hospital there, they delivered the medicine and several tons of baby formula that had arrived earlier. Then five members of the group boarded a smaller craft and headed to Narovlia and Bragin, half-evacuated towns at the edge of the Dead Zone.

Later Michael recorded the adventure in his journal:

> *Nosediving over fields and trees and small towns, at 200 kilometers per hour, and then soaring into the clouds with this thrill-seeking pilot—it was better than any E-ride at Disneyland. We sat unstrapped on a bench opposite a huge fuel tank with cotton stuffed in our ears to muffle the sound.*
>
> *I couldn't help but focus on the irony of all this. Here we were, American volunteers and former enemies, in a military machine once used in the Afghan War, delivering life-saving medicine and humanitarian assistance at the edge of Chernobyl's Dead Zone. We really were on a wild frontier of international relations.*
>
> *Valeri, a member of the military crew, said through the interpreter, "I was in a helicopter like this during the Afghan War when we were shot by a U.S. Stinger missile. The pilot died and I parachuted to safety." After a pause, he added, "I'm glad we can now use this helicopter for a mission to save lives rather than to end lives."*

❊ ❊ ❊

SITTING ON A GRASSY, tree-lined bank above the Svisloch River, the little green cottage at No. 1 Communisteecheskaya Street in Minsk was the reconstructed site of the First Congress of the Social Democratic Workers Party in 1898. In the original cottage (which was destroyed during the Great Patriotic War), within sight of a czarist militia post, Communists set out to create a new world.

Up until fall 1991, the green cottage—known as the Cradle of Communism and owned by the Communist party of

Belarus—had been a "must" stop on every Intourist-guided tour of the city; then, in an instant, it was confiscated, along with other party property, and boarded up. It was an embarrassment to the new democrats who were determined to blot out a painful past.

In early 1992, not long after the Soviet Union had been officially dissolved, Paul Moore and Volodya, an interpreter, left the Byelorussian Children's Fund office at No. 2 Communisteecheskaya (on the ground floor of the building that had housed Lee Harvey Oswald during his Minsk years). They strolled to the cottage across the street and rapped on the door. The curator was inside. Although the museum was closed, he agreed to show them around.

With great care and pride, he pointed out the Communist party archives and artifacts, identifying familiar faces in photographs on the walls. He was not the least bit embarrassed by his museum, only by the low esteem in which it was then held.

Remarkably, the curator had a Russian Bible on his desk, and Paul picked it up. He handed it to Volodya and asked him to read Acts 2:44-45 aloud in Russian:

> All who believed were together and had all things in common; they would sell their possessions and goods and distribute the proceeds to all, as any had need.

"Is this what the men at the First Congress had in mind as they dreamed about a new world?" Paul asked the curator, who was astonished that an ideal that communists held so dear could be found in the Bible.

Like Paul the apostle, who had commended the Athenians for building an altar "To an unknown God," Paul Moore was quick to see the possibilities in this link between socialist history and the Christian message. He promptly launched one of his grand, outrageous schemes: Why not preserve the museum in the rooms on the left (where else?) side of the house, and then rent the rest of the house to Citihope for its permanent headquarters in Minsk? The irony of the project positively delighted him.

Paul posed the notion to Pyotr Kravchanka, the Belarusan foreign minister, who tried to discourage him, suggesting that the move might be "misunderstood by the people." But Paul was characteristically tenacious, and Kravchanka finally agreed to be Paul's advocate with the state.

After hearing about Paul's plan, Michael Christensen turned to his friend Sasha Lukashuk, whose readings on local issues Michael trusted. "I already know that you hate everything that building stands for. What do you think of the idea?"

"Impossible! It will never happen. They will not allow it."

But he was wrong. Within weeks of the proposal, while the Operation *Nadezhda* Express volunteers were still in the city, a preliminary lease was drawn up with the understanding that Citihope International would be registered as a charity under the laws of Belarus. The legalities were met and the Moores opened the newly painted rooms on the right side of the cottage as headquarters for Hope for Belarus, Citihope's official name in Minsk.

That evening, Michael again asked Sasha what he thought of the bold reopening of the little green cottage in this way.

"What an amazing thing to redeem this building like this," he marveled. "I must tell you a story. Last winter, I was at a party at the Parliament building. After three vodkas, I announced a little too loudly that I would give a thousand rubles to anyone who would burn the cottage down. Later, as we walked home, I upped the wager to two thousand rubles.

"Now, to think that this museum of communism will have a humanitarian and religious purpose. It's really quite remarkable."

But that wasn't quite the end of the story, as Sasha Lukashuk informed Michelle by fax. A short time after the Citihope team departed at the end of Operation *Nadezhda* Express, attention again focused on the little green cottage.

"Communists would not be communists if they allowed such a chance for propaganda to slip away. Only a month before, a 'new' Communist party had been registered in Belarus in place of the old one, which had been banned after the coup," Sasha explained. "The museum was chosen for this test of wills. Posters appeared all around—'Yankee, go home!' 'Minister of Culture, resign!' and 'Give us back our history!'

"Activists started to collect signatures in support of their demands; pickets told rare passersby that 'those mean American capitalists have bought up our glorious past as well as the museum.' I don't mind telling you that I wish someone really had!

"In fact, there were barely two or three dozen of those fanatics and *apparatchiks*, but they knew how to make themselves heard. The ears of those in authority in Belarus are still tuned to the Communist wavelength, and on the fourth or fifth day of the protest, the Minister of Culture agreed to cancel the Citihope contract and reopen the museum unchanged."

The issue was finally resolved when all parties agreed to a permanent exhibition in one room of the museum about the children of Chernobyl and the work of Citihope International. Citihope agreed to move its offices to the building which then housed the offices of the Commonwealth of Independent States.

"I told this story about the Americans 'buying' our culture to an old veteran I met at Komarovka market one day," Sasha wrote. "He became quite agitated. 'The Communists have been robbing our country for the past 70 years and have robbed enough for themselves to last another 70 years. They don't need any help from America, do they?' Even the market salesmen, themselves experts in strong language, listened with respectful amazement to his string of choice wishes for the Communist party."

❊ ❊ ❊

"IT ALL BEGAN in that little green cottage," Sasha said.

Michael had asked him about Kurapaty, Stalin's killing field outside Minsk where the remains of as many as a quarter of a million victims may lie.

"It began there with men who wanted to create a new world. In order to make a new world, they had to make a new people. They had to propose new ideas and get rid of old ideas, get rid of old thinking and, finally, get rid of the people who were old-thinking.

"Repression and propaganda gave way to punishment and elimination," Sasha recalled in a quiet voice.

"First the priests were arrested and made to confess their treason against the state. Once they signed confessions, they were taken out at night into the woods and shot. Next came the poets and artists and intelligentsia. They, too, died in the night. Finally, the rank and file.

"So far, 508 mass gravesites, which were filled from 1937 to 1941, have been found. The systematic mass murders were committed in cold and calculated fashion in the deep of the night.

Bodies lay on top of bodies, layer upon layer, until each grave was filled.

"Rumors of the graves surfaced from time to time but, until the late 1980s, no one dared to believe," Sasha explained. Then in 1988, after ten years of very private interviews and excavations, Zenon Posniak, an archaeologist and leader of the Belarusan Popular Front, published the first shocking article on Stalin's slaughter at Kurapaty.

Slowly the dimensions of Kurapaty's tragedy began to be revealed. As disclosure after disclosure was made, the Popular Front created a shrine at Kurapaty that became a rallying site for the political opposition.

"People like to come and leave flowers at the cross and plant pine trees on the anonymous graves," Sasha said. "The KGB says there are no documents left. We still do not know the name of a single victim of Kurapaty."

Sasha's story captivated Michael, and he was determined to see Kurapaty himself. The next day a trip to Khatyn, the memorial to the destroyed villages of Belarus, was scheduled for the volunteers. After the tour, Michael pressured the guide to take the group to Kurapaty, but she gave one excuse after another for not going: there wasn't enough time; the site was under excavation; she wasn't sure where it was.

So Michael rented the red Intourist bus again the following day and the volunteers piled back on. Armed with directions from Sasha, Michael led the group to the site just outside the city, off Ring Road. An unpaved road—really just a country lane—rambled through the birch trees to the edge of a depressed clearing, sort of a forest canyon.

For more than an hour, the Americans walked silently among the graves. They stood for a while at the "People's Monument," a rock with a chiseled inscription erected by the new government as a promise to build a more formal and permanent memorial on the site. Then they moved down the trail to a small wooden cross that Poles had put up in memory of the large number of their own people who perished there.

Finally, they found the large, rough-hewn wooden cross near the highway with its barbed wire "crown of thorns" and hand-lettered inscription beginning, "1937-1941." In silence, they lit candles in the sand at the base of the cross and prayed. Michael led them in a ceremony of remembrance at the gravesite, with special mention of the father of Vadim Karpenko, one of the volunteers, whose father had been killed by Stalin.

"Kurapaty, Khatyn, Chernobyl—the three Golgothas of Belarus," Sasha had told Michael. "We know that Christ was resurrected after his Golgotha, but no one knows if this nation can be resurrected after three Golgothas."

❋ ❋ ❋

SOME UNFINISHED BUSINESS awaited the three from the Congregational Church of Belmont who had carried seven cartons of medicine with them from Kennedy Airport in New York.

Dr. Olga's appeals for help had drawn Michelle back to her hospital three times since their first meeting in November 1990. Each time she had arrived with medicines on Dr. Olga's "shopping list" that Paul Maxey kept current by fax communication. And each time she observed dramatic improvements in the hospital from the previous visit.

At the same time, Dr. Olga had made some disturbing discoveries. "We are noticing far more cases of families with more than one child with leukemia. Right now, we have five families, each with two children sick. In all my nineteen years as a doctor, I know of only one other case of two in one family, and that was a mother and a daughter. Now we have three sets of siblings and two sets of cousins.

"This, we believe, is connected to radiation from Chernobyl."

Another development was a much younger onset of leukemia. "In the West—and here before Chernobyl—the youngest children we usually saw were five to seven years old. Now we are seeing children as young as two to four years old with leukemia."

Dr. Olga reiterated her commitment to treating all the children of Chernobyl, even the sickest, right there in Belarus. "We must do it ourselves, even bone-marrow transplantation. Too few children can be treated if you take them to the West one at a time."

She reported that her dream of a children's cancer center in Minsk was moving closer to reality. The land had been set aside and "sixteen million of the thirty million dollars needed has been raised in Germany, Switzerland, Austria, and Japan."

She saved her truly spectacular news for last. "I just got back from a conference in Vienna, where I presented a paper," she said as she pulled some slides of bar graphs from a folder. "Look at this. For 158 recent cases, the percentage of children going into first remission—eighty-five percent!"

Her eyes flashed with satisfaction, and Michelle remembered how she had once challenged them to "send me methotrexate if you want to help. Then my children will live, eighty-five percent of them, just as in the West!"

Dr. Olga's miracle was definitely taking shape.

EPILOGUE

Just two years after Pyotr Kravchanka had quoted the book of Revelation to the General Assembly of the United Nations, the new nation of Belarus had become the focus of a large-scale, international effort to embrace and serve the children of Chernobyl.

Dr. Olga Aleinikova's once-threadbare hospital was now bustling with computerized equipment and blue gingham linens; eighty-five percent of her children were living instead of dying. Vital pharmaceuticals such as methotrexate were flowing somewhat consistently into the Children's Hematological Center, while earthmovers prepared ground for a new multidisciplinary children's cancer center.

Homes and summer camps across the United States were offering radiation-free rest and recreation, while American and European mental health professionals prepared to open a clinic in the contaminated region.

As the people of Belarus lived with the tragic effects of their three Golgothas—Khatyn, Kurapaty, and Chernobyl—they were still facing a bleak future full of seemingly insurmountable challenges. Yet a personal covenant had been forged with the Western volunteers who were regularly bringing medicine, food, care, and concern—a covenant that was raising hope from the ashes of despair.

LIST OF ORGANIZATIONS

After reading this book you may be wondering how you can contribute to relief efforts on behalf of the children of Chernobyl. Here is a list of organizations that would welcome your interest. You may already have provided some help without knowing it. A portion of our authors' royalties will be used to buy lifesaving medicine and mental health services for the children you have read about.

Chernobyl Children's Project
Radiation Break
Attn: Connie McClain
340 King Rd
Petaluma CA 94952

Children of Chernobyl Fund
Allied Medical Ministries Inc
24 Windsor Rd, Suite 102
Westminster MD 21157-4413

Children of Chernobyl Project/Medicine
Attn: Michelle Carter
Congregational Church of Belmont
751 Alameda de las Pulgas
Belmont CA 94002

Children of Chernobyl Project/Mental Health Service
Attn: Michael Christensen
1032 Irving St, Box #230
San Francisco CA 94122

Citihope International
Attn: Paul Moore
PO Box 38
Andes NY 13731-7900

Friends to Friends
Attn: Nick Blea
8708 La Sala Grande NE
Albuquerque NM 87111

Muskegon YFCA Byelorussian Fund for Children
Attn: Mary Beth Ormiston
900 West Western
Muskegon MI 49441

World Vision Children of Chernobyl Project
Attn: Serge Duss
919 West Huntington Dr
Monrovia CA 91016

GLOSSARY OF RUSSIAN WORDS

absolutva (absolutely)
Amerikanka (female American)
apparatchik (high-level Communist party officials)
avtobus (bus)
babushka (a grandmother or old woman)
boyar (rich merchant)
chernobyl (bitter ground)
chut, chut (just a little bit)
dacha (summer cottage)
documentee (documents)
dukhovnost (spirit)
dyedushka (grandfather)
dzhernaya (floor clerk at a hotel)
electreechka (electric train)
gazyeta (newspaper)
glasnost (the new spirit of openness in USSR under
          Gorbachev)
gulag (prison camp)
khleb (bread)
kopeck (100 kopecks equals 1 ruble)
krasnaya zvezda (red star symbolizing Soviet state)
medicamentee (medicine and medical supplies)
menya zavoot (my name is)
nadezhda (hope)
nevyezdnoi (person deemed too unreliable to be permitted
          to travel abroad)
nomenklatura (Communist party elite)
nyet (no)
oblast (a local political subdivision, equivalent to a county in
          the U.S.)

perestroika   (economic liberalization in the USSR under
     Gorbachev)
piroshkee   (a meat pie)
produktee   (food store)
rozhestvo   (Christmas)
ruble   (basic unit of money in USSR)
spaseebuh   (thank you)
stolovaya   (dining room)
svyatoi   (saint or holy one)
"Vremya"   (a Soviet nightly news program; literally, "The
     Times")
Zorachka   (Belarusan proper name meaning "little star")